[英国] 萨米尔·奥卡沙 著　韩广忠 译

牛津通识读本·

科学哲学

Philosophy of Science

A Very Short Introduction

译林出版社

图书在版编目（CIP）数据

科学哲学／（英）奥卡沙（Okasha, S.）著；韩广忠译. —南京：
译林出版社，2013.6（2024.11重印）
（牛津通识读本）
书名原文：Philosophy of Science: A Very Short Introduction
ISBN 978-7-5447-3276-5

Ⅰ.①科…　Ⅱ.①奥…②韩…　Ⅲ.科学哲学　Ⅳ.①N02

中国版本图书馆 CIP 数据核字（2012）第 219851 号

Philosophy of Science: A Very Short Introduction By Samir Okasha
Copyright © Samir Okasha 2002

著作权合同登记号　图字：10-2014-197 号

科学哲学　〔英国〕萨米尔·奥卡沙 ／ 著　韩广忠 ／ 译

责任编辑　许　丹　何本国
责任印制　董　虎

原文出版　Oxford University Press, 2002
出版发行　译林出版社
地　　址　南京市湖南路 1 号 A 楼
邮　　箱　yilin@yilin.com
网　　址　www.yilin.com
市场热线　025-86633278
排　　版　南京展望文化发展有限公司
印　　刷　南通印刷总厂有限公司
开　　本　890 毫米 × 1260 毫米　1/32
印　　张　9.125
插　　页　4
版　　次　2013 年 6 月第 1 版
印　　次　2024 年 11 月第 20 次印刷
书　　号　ISBN 978-7-5447-3276-5
定　　价　39.00 元

序言

李醒民

从 1978 年读研究生算起，我接触、关注、研究科学哲学（philosophy of science）已有三十个年头了。文章和著作没有少撰写，译文和译著没有少迻译，阅读的科学哲学书籍当然不会太少。但是，像手头这本通俗易懂、言简意赅、语句流畅的《科学哲学》一书，我着实是第一次谋面。

这自然是我浏览了译者译稿之后才有的感受。不过，从作者为本书所取的名字——直译则为《科学哲学：非常简明导论》（*PHILOSOPHY OF SCIENCE：A Very Short Introduction*）就可一眼看出，这个副标题可谓名副其实：该书仅有短短的七章，篇幅至多十万余字；作者萨米尔·奥卡沙（Samir Okasha）教授论述深入浅出、行文平实，把相当深奥的哲学道理讲得条理分明、头头是道——这是一般人很难做到的，因而显得尤为难能可贵。我佩服作者这种大处着眼、小处落笔，化奥旨为直白的写作风格。作为"牛津通识读本"之一出版，这本小书是当之无愧的。对于中国广大科学哲学爱好者来说，它的确是一本别开生面的速成入门读物。读者借以曲径通幽，也许能够对科学哲学的良辰美景略窥一二。说不定由此生发开来，养成浓厚的兴趣，还会进一步探赜索隐、钩深致远呢。

读者切不要产生错觉，以为书小内容就少，通俗就意味着浅薄。绝非如此！几年前，我在为《中国科学哲学论丛》(李醒民、程承斌主编)所写的新序中指明：科学哲学的研究范围和边界虽然难以精确划定，但是依然可以大致勾勒它的四个论域或内涵。PS1即科学哲学元论。它涉及科学哲学的根本性论题，是科学哲学的"形而上学"层次，与科学知识本身相距较远。例如，科学的目的、目标、对象、价值、范围、限度、划界、方法、预设、信念等等。PS2即科学哲学通论。它涉及科学哲学的普遍性论题，与科学知识整体的关系密切。例如，科学的问题、事实、概念、定律、原理、理论结构，科学的发现和发明、证明和辩护、说明和诠释、语言和隐喻，科学的发展、进步、革命，科学中的机械论和有机论、还原论和活力论、进化论和目的论、因果性和几率性、连续性和分立性，对科学的经验论、理性论、实在论、反实在论、现象论、工具论、整体论、操作论、约定论、物理主义、历史主义、后现代主义的解读等等。PS3即科学哲学个论。它是科学各门分支学科中的哲学问题。例如物理学、生物学、系统论、信息论、复杂性科学等中的哲学问题。如果说以上三个论域大体属于科学哲学内论的话，那么PS4则可以称为科学哲学外论。它的主要研究对象是科学活动和科学建制的本性以及科学与外部世界——自然界、社会、人——的错综复杂的关系。例如，科学的规范结构和精神气质，科学的起源，科学的社会文化功能，科学与人生和人的价值，科学与政治、经济、文化、艺术、哲学、伦理、宗教的内在关联和外在互动等等。(李醒民：《科学哲学的论域、沿革和未来》)打开这本《科学哲学》，我们看到它涉及的论题相当广泛：什么是科学，科学推理，科学中的说明，实在论和反实在论，科学变化和科学革命，物理学、生物学和心理学中的哲学问题,科学和科学非难者。两相对照,读者不难窥见,这本小书或多或少涵盖了科学哲学的四个论域。如果仔细阅读一下它的具体内容,读者不难对此有更为切身的体察。作者善于将大义寓于微言,苟非工夫积久,博观而约取,厚积而薄发,焉能成竹在胸、举重若轻?

众所周知,当今之世,往往被称为科学时代。科学在社会中的地位和重要性,由此略见一斑。单说科学,它以技术——科学的副产品和衍生物——为中介可以转化为无与伦比的物质力量,使我们的衣食住行发生翻天覆地的变化,给人类带来史无前例的福利。这一点有目共睹,毋庸赘言。不仅如此,科学本身作为人类伟大的思想创造和文化成就,也具有震撼人心的精神力量;也就是说,科学具有深邃幽远的精神价值(*李醒民:《论科学的精神价值》*)或精神功能(*李醒民:《论科学的精神功能》*)。例如,破除迷信和教条的批判功能,帮助解决社会问题的社会功能,促进社会民主、自由的政治功能,塑造世界观和智力氛围的文化功能,认识自然界和人本身的认知功能,提供解决问题的方法和思维方式的方法功能,给人以美感和美的愉悦的审美功能,训练人的心智和提升人的思想境界的教育功能,如此等等不一而足。遗憾的是,人们往往看不到这一点——实无异于有眼不识泰山。

在现代社会,凡是具有中学文化程度的人,都多少具有一些普通的科学知识。随着大学教育的扩大和普及,具有专门科学知识的人越来越多。令人扼腕叹息的是,人们了解和把握的只是科学知识,人们窥见和注重的只是科学的物质成就。他们既不了解作为一个整体的科学的丰富内涵(作为知识体系、研究活动和社会建制的科学),也不把握科学的精神底蕴和文化意义，当然就更不知道如何以公允的态度和平和的心态正确看待科学了。这种无知状况不仅遍布于普通人，而且在知识分子、社会精英、政府官员中也不乏其人，乃至科学共同体的诸多成员亦"不识庐山真面目"①。科学哲学作为对科学进行反思和批判的哲学学科，正是探讨这些问题的。因此，为了了解和把握科学，有必要学点科学哲学。要知道，在现时代，科学已经成为社会的中轴，科学文化已经成为人类文化的特采，而且正在铸造世界的未来。在这样的科学时代，不懂一点科学知识，肯定不能算是现代人；而一点不懂科学哲学，恐怕也难以步入现代人的行列吧？由此看来，奥卡沙教授的小书不啻雪中送炭。对于需要进一步深究的读者，不妨找一些西方科学哲学和科学文化的经典著作读读。在这方面，拙作《科学的文化意蕴——科学文化讲座》和《科学论：科学的三维世界》②，也许能起点锦上添花的效用。

　　作为一本科学哲学普及书籍，《科学哲学》得以翻译、出版，是很有意义的。我希望它能惠及中国的广大读者，对于提高国人的科学素养起到促进作用。当读者洞悉玄奥于涣然冰释之时，指点迷津于山重水复之处，何尝不是一件快事和雅趣？是为序。

----

　　① 我们的这一估计是切合中国实际的。即使在科学发达的西方国家，情况也与此类似。这里有威兹德姆的言论佐证："科学时代的任何一个人，几乎不知道科学的本性是什么。这不仅包括那些通过周刊意识到科学的人，而且也包括哲学家和科学家本身在内。" 参见 J. O. Wisdom, *Challengeability in Modern Science*, Avebury, 1987, p. 16.

　　② 李醒民：《科学的文化意蕴——科学文化讲座》，北京：高等教育出版社，2007 年 5 月第 1 版。《科学论：科学的三维世界》已经基本完成，正在联系出版社出版。

目录

致谢

比尔·牛顿－史密斯、彼得·利普顿、伊丽莎白·奥卡沙、利兹·理查森、谢莉·考克斯诸君阅读本书初稿并提出了意见，谨致谢意。

萨米尔·奥卡沙

何为科学？

什么是科学？这个问题似乎很容易回答：每个人都知道科学包含诸如物理、化学和生物等学科，而不包括艺术、音乐和神学之类的学科。但是当我们以哲学家的身份询问科学是什么的时候，上述回答就不是我们想要的那种回答了。此时我们所寻求的不是一个通常被称为"科学"的那些活动的清单，而是清单上所列学科的共同特征，换言之，**使科学得以**成为科学的东西是什么。这样一来，我们的问题就不再显得那么平凡了。

但是，也许你仍然认为这个问题有些简单化。科学真的只是在试图理解、解释和预言我们生活于其中的世界吗？这当然是一种合理的答案。但是仅仅如此吗？毕竟，各种宗教也同样在试图去理解和解释世界，可是通常并不被看做科学的一个分支。同样地，虽然占星术和算命也在试图预言未来，但大多数人并不将这些活动称为科学。再来考虑一下历史。虽然历史学家的目的是理解和解释过去发生的事件，但是历史通常被归为人文学科而不是科学学科。和许多哲学问题一样，"何为科学？"这个问题实际上比初看上去难解得多。

许多人认为科学的显著特征在于科学家探索世界的特殊方法。这种观点似乎相当有理。因为许多科学的确使用了在其他非科学的学科中找不到的特殊方法。一个明显的例子就是实验方法的运用,它是现代科学发展史上的转折点。然而,并不是所有的科学都运用实验方法——天文学家显然不能在天上做实验,有时必须代之以仔细的观察。在许多社会科学领域,情形也是如此。科学的另一个重要特征是科学理论的建构。科学家并不是仅仅在记录簿上记下他们实验和观察的结果——他们通常希望用一个一般的理论来解释那些结果。虽然这并不是总能很轻易地做到,但已经获得了一些重大的成果。科学哲学的一个关键问题就是去弄明白实验、观察和理论建构等方法是如何帮助科学家揭开那么多自然之谜的。

现代科学之起源

在今天的中小学和大学里,基本是以非历史的方式来教授科学的。教科书采用尽可能方便的形式来表述科学学科的关键思想,很少涉及促成这些科学发现的漫长而又经常曲折发展的历史过程。作为教学方法,这样做是有道理的。但是对于科学思想发展史的适当关注会对理解科学哲学家感兴趣的那些论题有所助益。实际上,我们将在第五章看到得到论证的这种观点:对科学史的密切关注是做好科学哲学工作所必不可少的。

现代科学起源于 1500 年到 1750 年之间发生在欧洲的科学高速发展时期,即我们现在所称的科学革命时期。当然,古代和中世纪的人们也从事科学探索——科学革命并不是凭空产生的。在这些早期阶段,主流的世界观是亚里士多德学说,这一名称来自古希腊哲学家亚里士多德。亚里士多德在物理学、生物学、天文学和宇宙学领域都提出了具体的理论。但是正如他的研究方法那样,亚里士多德的观点对于一个现代科学家来说似乎是非常古怪的。仅举一例：他认为所有的地球物体都仅是由土、火、空气和水四种物质组成的。这种观点显然与现代化学告诉我们的东西相冲突。

在现代科学世界观的发展过程中，第一个关键阶段是哥白尼革命。1542 年，波兰天文学家尼古拉斯·哥白尼（1473—1543）发表了一本抨击地心说宇宙模型的著作，地心说模型认为静止不动的地球位于宇宙的中心，行星和太阳都在围绕地球的轨道上旋转。地心说式的天文学也称为托勒密天文学，以古希腊天文学家托勒密的名字命名。它是亚里士多德式世界观的核心，延续了约 1800 年而未受质疑。但是哥白尼却提出了另外一种观点：**太阳**是宇宙的固定中心，包括地球在内的行星都在环绕太阳的轨道上运行（参见图 1）。在这种太阳中心说的模型中，地球仅被看做是另外一个行星，因此也就失去了传统曾经赋予它的独特地位。哥白尼的理论最初遇到了非常大的抗拒，尤其是来自天主教会的抗拒。天主教会认为哥白尼的理论是对圣经的背叛，并于 1616 年禁止了宣扬地动学说的书籍的发行。然而在不到 100 年的时间里，哥白尼学派就被确立为正统的科学。

哥白尼的革新不仅带来了更好的天文学的进步，通过约翰内斯·开普勒（1571—1630）和伽利略·伽利雷（1564—1642）的努力，它还间接地推动了现代物理学的发展。开普勒发现，行星围绕太阳运行的轨道不是哥白尼所猜想的正圆形，而是椭圆形。这就是他重要的行星运动"第一定律"；他的第二和第三定律明确给出了行星围绕太阳运行的速度。

开普勒的三定律加在一起，给出了一个远比以前提出的理论更好的行星运动理论，解决了许多世纪以来困扰天文学家的难题。伽利略终生追随哥白尼的学说，也是望远镜的早期发明

图 1 哥白尼的日心说宇宙模型，描绘了包括地球在内的行星围绕太阳
旋转的情形。

人之一。当把望远镜对准天空的时候，他得到了许多惊人的发现，其中包括月亮上的山脉、大量的恒星、太阳黑子以及木星的卫星。所有的这些发现同亚里士多德学派的宇宙学完全相矛盾，并在科学共同体转向哥白尼学说的过程中发挥了至关重要的作用。

然而，伽利略最持久的影响并不在天文学，而是在力学中；他推翻了亚里士多德学说中关于重物体比轻物体下落速度更快的理论。取而代之的是，伽利略提出了一种反直觉的观点，认为所有做自由落体运动的物体都以相同速率向地面下落，不受重量影响（参见图 2）。（当然，在实践中如果你从相同的高度向下抛一片羽毛和一枚炮弹，炮弹将会首先着地，然而伽利略认为这仅仅是由于空气阻力的作用——在真空中,它们将会同时着地。)另外,他还认为做自由落体运动的物体是均匀加速的,即在相等的时间内获得相等的速度增量;这就是伽利略自由落体定律。伽利略为这一定律提供了尽管不是决定性的但却具有说服力的确凿证据,这构成了他力学理论的核心部分。

通常认为,伽利略是第一位真正的现代物理学家。他第一次表明数学语言可被用来描述物质世界中的真实物体的行为,例如下落的物体、抛射的物体等等。在我们看来这似乎是很显然的——今天用数学语言来表述科学理论已经成为惯例,不仅是物理学,在生物学以及经济学领域也是如此。但在伽利略的时代,这却不是显然的:人们普遍认为数学处理的是纯粹抽象的实体,因此对于物质实体是不适用的。伽利略所做工作的另外一个革新方面是,他强调了运用实验来检验假说的重要性。对于现代科学家来说,这也许又是一个看上去显而易见的观点。但是,在伽利略的时代,人们并不认为实验是一种获得知识的可靠手段。伽利略对于实验检验的强调标志着一种研究自然界的经验方法的出现,这一方法一直沿用至今。

伽利略去世后接下来的那段时期,科学革命突飞猛进。法国哲学家、数学家和科学家勒内·笛卡尔(1596—1650)提出了一门全新的"机械论哲学",按照这种哲学,物理世界仅由相互作用和相互碰撞的惰性粒子物质构成。控制这些粒子或"微粒"运动的定律就是理解哥白尼式宇宙结构的关键因素,笛卡尔对此深信不疑。机械论哲学声称将用这些惰性的、不可感知的微粒运动来解释一切可观察现象,很快就成为了17世纪下半叶的主流科学观;在某种程度上至今它仍然影响着我们。机械论哲学的观点得到了诸如惠更斯、伽桑狄、胡克、玻意耳等人的支

'They were seen to fall evenly.'

图 2　素描：伽利略测量从比萨斜塔落下物体的速度的神奇实验。

持;对它的广泛接受标志着亚里士多德式世界观寿终正寝。

科学革命在艾萨克·牛顿(1643—1727)的研究工作的推动下达到了顶峰,他的贡献在科学史上无人可出其右。牛顿最杰出的著作是《自然哲学的数学原理》一书,出版于 1687 年。牛顿虽然赞同机械论哲学家们关于宇宙完全是由运动粒子构成的观点,但他却试图改进笛卡尔运动定律和碰撞规则。其结果是,在牛顿的三大运动定律和他著名的万有引力原理的基础之上,强大的动力学和机械论理论诞生了。按照该定律,宇宙中的每一个物体都对所有其他物体产生引力;两物体间引力的大小取决于它们质量的乘积和它们之间距离的平方。运动定律阐明了引力是如何影响物体运动的。牛顿发明了今天被我们称为"微积分"的数学技巧,对理论的表述具有很高的数学上的精确性和严格性。令人惊奇的是,牛顿能够表明开普勒的行星运动定律和伽利略的自由落体定律(经过微小的修正)都是他的运动定律和万有引力原理的逻辑结果。换言之,无论是天上的还是地上的物体运动,都可以用同样的定律来解释。牛顿给出了这些定律精确的定量形式。

牛顿物理学为此后 200 年左右的科学提供了框架,很快就取代了笛卡尔物理学。主要由于牛顿理论的成功,科学的信心在此期间迅速增强。人们普遍认为牛顿的理论揭示了自然界真正的运行方式,并能够解释一切,至少在原则上是可以的。人们作了更为细化的尝试,以便把牛顿力学的解释模式拓展到越来越多的自然现象上。18 和 19 两个世纪见证了巨大的科学进步,尤其是在化学、光学、能源、热力学以及电磁学研究领域。但是大多数情况下,这些新发展都被看做是在一个宽泛的牛顿宇宙观范围之内作出的。科学家们把牛顿的观念作为最根本的正确观念来接受；剩下的工作就是在细节上对其加以填充而已。

牛顿式的理论图景在 20 世纪上半叶被动摇了，这要归功于物理学上两项革命性的新发展：相对论和量子力学。爱因斯坦发现的相对论表明，在运用于特别巨大的物体或者运动速度极快的物体时，牛顿力学无法给出正确的解答。相反的是，量子力学则指出在运用于微观领域的亚原子微粒时，牛顿力学无法给出正确解答。相对论和量子力学两者，特别是后者，是非常奇特和激进的理论，它们关于实在本性的论断使很多人难以接受甚至难以理解。它们的出现导致了物理学上重大的观念变革，这些变革一直延续至今。

到现在为止，我们对于科学历史的简要回顾主要集中在物理学领域。这绝非偶然，物理学不仅在历史上非常重要，在某种意义上也是所有科学学科当中最基础的学科。这是因为，其他科学的研究对象本身都是由物理实体构成的。以植物学为例。植物学家研究植物，植物最终是由分子和原子构成的，这些分子和原子都是物理学微粒。因此，植物学显然不如物理学更基础——尽管这并不是说它更不重要。我们将在第三章回到这一点进行讨论。但是，如果完全忽略非物理科学，对现代科学起源的一个即使是简要的阐述也将是不完整的。

在生物学领域，最著名的事件是查尔斯·达尔文关于通过自然选择实现物种进化的理论发现，这一理论 1859 年被发表在《物种起源》一书中。在此之前，按照圣经《创世记》的教导，人们普遍认为不同的物种都是由上帝分别创造的。但是达尔文却认为，当代的物种事实上都是由古代的物种通过一种名为自然选择的过程进化而来的。当一些生物组织依靠它们的本身特征比其他的组织留下更多的后代时，自然选择就开始了；如果

科学哲学

MR. BERGH TO THE RESCUE.

THE DEFRAUDED GORILLA. "That *Man* wants to claim my Pedigree. He says he is one of my Descendants."

Mr. BERGH. "Now, Mr. DARWIN, how could you insult him so?"

图 3 达尔文关于人类和大猩猩是从相同祖先演化而来的观点震惊了维多利亚时代的英格兰。(图中文字为：伯格先生来解围。受骗的猩猩：那个人想挤进我们的家谱。他说他是我的后代。伯格先生：哎呀，达尔文先生，你怎么可以那样侮辱他。）

这些特征被它们的后代所继承，随着时间的推移，这一种群就会越来越好地适应环境。达尔文认为，尽管这一过程很简单，但是经过许多代之后，它就会导致一个物种进化成另一个全新的物种。达尔文为他的理论提供的证据非常有说服力，以至于在

图 4 詹姆斯·沃森和弗朗西斯·克里克以及他们于 1953 年发现的
DNA 结构的分子模型——著名的"双螺旋结构"。

20 世纪开始之前它就作为正统的科学被人们接受了，尽管有
许多来自神学的反对意见(参见图 3)。后续的科研工作为达尔
文的理论提供了更为惊人的验证,这一理论成为了现代生物学
世界观的核心观点。

　　20 世纪又见证了另外一场迄今尚未完成的生物学革命:分
子生物学、特别是分子基因学的问世。1953 年沃森和克里克发
现了生命体细胞当中组成基因的遗传物质 DNA 的结构（参见
图 4）。沃森和克里克的发现解释了基因信息如何从一个细胞

被复制到另一个细胞,并从父母传给子女的问题,从而解释了为何子女往往与父母相像。他们的发现开辟了生物学研究的一个激动人心的新领域。在沃森和克里克的发现问世以来的50年里,分子生物学获得飞速发展,改变了我们对遗传以及基因如何构建生物体的理解。最近试图进行的对人类体内整套基因提供分子水平描述图的工作,即人类基因组计划,标志着分子生物学的深远发展。21世纪将会见证这一领域更加激动人心的进步。

过去一百年间投入到科学研究方面的资源比以前任何时候都要多。带来的一个局面就是新科学学科,诸如计算机科学、人工智能、语言学和神经科学的大量涌现。也许近30年来最重要的事件就是认知科学的兴起,认知科学研究人类认知的各个方面,例如感知、记忆、学习和推理,并且改造了传统的心理学。认知科学很大的动力来自于一种观点,该观点认为人脑在某种程度上类似于一台计算机,人类的心智过程因而可以通过与计算机执行的操作加以对比得到理解。认知科学虽然仍处于婴儿期,但很有希望依靠它揭示关于意识活动的大量机理。社会科学,特别是经济学和社会学,在20世纪也得到了繁荣发展,尽管许多人认为它们在成熟性和严格性方面仍落后于自然科学。在第七章我们将回到这个问题上来。

何为科学哲学?

科学哲学的主要任务是去分析各门科学所采用的研究方法。你也许会疑惑,为何这一工作应该由哲学家承担,而不是科学家们自己呢?这是一个很好的问题。部分的回答是,从一个哲学化的视野来观察科学可以使我们进行更深入的探索——去揭示科学实践中暗含的但不被科学家们明确讨论的假设。以科学实验为例来作一解释:假设一个科学家做了一个实验并且获得了一个特定的结果。他反复多次做这一实验,一直得出相同的结果。然后他可能会停下来,并相信如果他继续在完全相同的条件下做这一实验,得到的结果将一直相同。这一假设也许看起来很显然,但是作为哲学家,就会质疑。有何理由让我们假设将来的重复实验会得到相同的结果?我们怎么知道这是真的呢?科学家不可能花费太多的时间来厘清这些略显古怪的问题:他也许有更好的事情去做。这些是纯粹的哲学问题,我们在下一章会加以阐释。

因此,科学哲学的部分工作就是去质疑科学家认为理所当然的假设。但如果我们暗示科学家自己从来都不讨论哲学问题,就有失偏颇了。事实上,历史上许多科学家在科学哲学的发展过程中发挥了重要的作用。笛卡尔、牛顿和爱因斯坦就是著名的例子。他们每一位都对一些哲学问题深感兴趣,这些问题包括科学应该如何进步,科学应该使用什么样的研究方法,我们对这些方法的信任度有多高,科学的认知是否有限度,等等。我们将会看到,这些问题仍然处于当代科学哲学的核心地带。所以,引起科学哲学家兴趣的论题并不是"纯粹哲学的";相反,它们曾经引起了历史上一些伟大科学家的关注。另一方面,我们也必须承认今天的许多科学家对科学哲学不感兴趣并且也对其缺乏了解。这当然不是好事,但这并不表明哲学问题已失去意义。倒不如说,这是自然科学加速专业化和现代教育体系中科学和人文学科两极分化的一个结果。

你现在也许仍然想确切知道科学哲学到底是什么。正如上文所说的,说它是"研究科学方法"的学问并没有交代得很确切。与其提供一个内容更加丰富的定义,我们不如通过直接考察科学哲学中的一个典型问题来进行解释。

科学和伪科学

回顾我们开始时提出的问题:什么是科学?作为 20 世纪一位颇有影响的科学哲学家,卡尔·波普尔认为科学理论的基本特征是它应具有可证伪性。称一个理论是可证伪的并不是说它是错的。而是说,它意味着该理论能够作出一些可以用经验进行检验的特定预测。如果这些预测被发现是错误的,这一理论就被证伪了,或者说被否证了。因此一个可证伪的理论是指我们能够发现它是错的——它不能和每一个可能的经验过程相容。波普尔认为一些所谓的科学理论是不满足这一条件的,因此根本不应该被称为科学;它们不过是伪科学。

弗洛伊德的精神分析理论是波普尔钟情的伪科学例子之一。按照波普尔的观点,弗洛伊德的理论可以与无论怎样的经验发现相一致。对于患者的任何行为,弗洛伊德学派都可以在他们的理论中找到针对性的解释——他们永远不会承认自己的理论是错误的。波普尔用下面的例子阐述了他的观点。设想一个带有蓄意谋杀倾向的人把一个小孩推到了河里,而另一个人为了救这个小孩牺牲了生命。弗洛伊德学派能够以同样轻易的方式解释两个人的行为:前者精神抑郁,后者已经获得了精神的升华。波普尔认为通过使用诸如精神抑郁、精神高尚和无意识的需求等概念,弗洛伊德的理论可以同任何临床数据相兼容;因此它是不可被证伪的。

波普尔认为,马克思的历史理论也存在这种问题。马克思主张,在全世界的工业化社会中,资本主义将让位于社会主义并最终走向共产主义。但是当这一断言没有变成现实之时，马克思主义者并没有承认马克思的理论是错的，而是会提出一种特殊的辩解来说明发生的事实现象其实与他们的理论完全一致。例如，他们也许会说走向共产主义的必然进程由于福利国家的兴起暂时减缓了速度，福利国家的兴起"软化"了无产阶级并削弱了他们的革命热情。采用这样的方法，马克思的理论就会变得同弗洛伊德的理论一样，与任何可能出现的事态相容。因此按照波普尔的标准，它们都不是真正的科学。

波普尔把弗洛伊德和马克思的理论同爱因斯坦的万有引力理论进行了比较，后者也被称为广义相对论。与弗洛伊德和马克思的理论不同，爱因斯坦的理论作出了一个非常明确的预测：来自遥远星球的光线会在太阳引力场的作用下发生偏转现象。通常这种效果是不会被观察到的——除非在日食的情况下。1919年英国天文学家亚瑟·爱丁顿爵士组织了两个探险考察队去观察那年的日食现象，一队去了巴西，另一队去了靠近非洲大西洋沿岸的普林西比岛，目的是验证爱因斯坦的预测。探险队发现星光确实被太阳偏折了，偏斜值几乎与爱因斯坦预测的完全一致。这件事给波普尔留下了极为深刻的印象。爱因斯坦的理论表达了一个确定的、精确的预测，这一预测被观察所证实。假如事实上星光没有被太阳偏折，就表明爱因斯坦是错误的。爱因斯坦的理论因此满足了可证伪性的条件。

波普尔将科学与伪科学区分开来的尝试直观上似乎是很合理的。一种可以符合任何经验数据的理论确实是值得怀疑的。但是有些哲学家认为波普尔的科学标准过于简单化了。波普尔批评弗洛伊德学派和马克思主义者通过解释来回避同他们的理论相矛盾的任何资料数据，而不是接受理论被推翻的事实。看上去他所批评的的确是一种值得怀疑的做法。但是，有证据表明这种做法被"有名望的"科学家们经常地采用——这些人并不是波普尔想要归入伪科学领域的科学家——并且已经带来了重要的科学发现。

另外一个天文学方面的例子可以解释这一点。上文提到的牛顿的万有引力理论预测了行星在围绕太阳旋转时应在的轨道。大多数情况下，这些预测通过观察得到了证实。然而，观测到的天王星的轨道却一直与牛顿的理论预测不一致。这一谜团在 1846 年被两位科学家揭开，英国的亚当斯和法国的勒威耶各自独立地完成了这一工作。他们认为存在着另外一个还未被发现的行星，它对天王星产生了附加的引力作用。假如该行星的引力作用的确是天王星轨迹异常的原因，亚当斯和勒威耶就能够计算出这颗行星应有的质量和应在的位置。不久，人们就在几乎恰好是亚当斯与勒威耶预测的位置发现了海王星。

现在很明显我们不应把亚当斯和勒威耶的行为斥为"非科学"——毕竟，这导向了他们对一颗新行星的发现。但是他们做的正是波普尔批评马克思主义者们所做的事情。他们开始于一个理论——对天王星轨道作出不正确预测的牛顿万有引力理论。他们没有断定牛顿的理论必定是错的，而是忠于这一理论并且试图通过假定存在一颗新行星的方式来解释与理论产生矛盾的观察事实。同样，当资本主义还没有显示出让位于共产主义的迹象的时候，马克思主义者没有得出结论说马克思的理论一定是错的，而是忠实于这一理论并且试图通过其他的方式来解释与理论相矛盾的观察现实。那么，如果我们承认亚当斯和勒威耶的探究方式是好的、的确是科学的范例，谴责马克思主义者是从事伪科学研究就确实不公平吗？

这就表明了波普尔将科学与伪科学区分开来的尝试是不完善的,尽管初看上去很有道理。亚当斯和勒威耶的例子绝不是个案。一般情况下,科学家们不会一遇到与观察数据相矛盾的情况就立即放弃他们的理论。通常,他们会寻找解决矛盾的方法而非放弃理论;这一点我们在第五章里还会谈到。值得牢记的是,事实上科学中的每一项理论都会和某一些事实现象相冲突——找到一个完全符合所有数据资料的理论是非常困难的。显然,如果一个理论与越来越多的数据资料一直相冲突,并且找不到解释冲突的合理方法,它最终将不得不被推翻。但是,如果科学家们在刚发现问题时就轻易抛弃理论,科学就不会有多少进步了。

波普尔所提出的科学标准的失败暴露了一个重要问题。是否真的能够找到所有被我们称为“科学”之物所共同具有且不被任何他物所拥有的特征呢?波普尔认为这个问题的答案是肯定的。他觉得弗洛伊德和马克思的理论显然是不科学的,因此必定会存在这些理论所不具有的而真正科学理论拥有的某些特征。但是,不管我们是否接受波普尔对弗洛伊德和马克思的否定评价,他关于科学拥有“本质特征”的设想都值得怀疑。毕竟,科学是一种多元性的活动,包含了范围广泛的不同学科和理论。也许它们共享一套能够定义何为科学的固定的特征,但也许这种特征并不存在。哲学家路德维希·维特根斯坦就认为,不存在能够定义何为“游戏”的一系列固定的特征;但却存在一束松散的特征,这些特征的大部分被大多数的游戏所拥有。然而也许某个特定的游戏不具有该特征束中的任一特征,却仍然是一个游戏。科学或许也是如此。如果真是这样,将科学与伪科学区分开来的一个简单化标准就不可能找到。

科学推理

    科学家们经常告诉我们一些关于世界的事实,这些事实如果不是出自他们之口,我们不会相信。例如,生物学家告诉我们,我们和大猩猩有密切的亲缘关系,地理学家告诉我们非洲和南美洲过去连接在一起,宇宙学家告诉我们宇宙一直在膨胀。但是,科学家们是如何获得这些听起来匪夷所思的结论的呢?毕竟,没有人曾经看到过一个物种进化成另一个物种,一块大陆分裂成两半,或者看到过宇宙变得越来越大。答案当然是,科学家们是通过推理或推论的过程确信上述事实的。对这种过程更多地了解对我们将会大有裨益。科学推理的确切本质是什么? 对于科学家们所作的推论我们应该持有多大的信任度? 这些就是本章所要讨论的话题。

演绎和归纳

    逻辑学家在演绎和归纳这两种推理形式之间作了重要的区分。下面是一个演绎推理或者演绎推论的例子:

所有的法国人都喜欢红葡萄酒
皮埃尔是一个法国人

因此,皮埃尔喜欢红葡萄酒

　　前两项陈述称为推论的前提,而第三项陈述称为结论。这是一个演绎推理,因为它具有以下特征:如果前提为真,那么结论一定也为真。换句话说,如果所有的法国人都喜欢红葡萄酒为真,并且皮埃尔是法国人也为真,那么就会得出皮埃尔确实喜欢红葡萄酒。这种推理通常可以表达为,推理的前提必然导致结论。当然, 这种推论的前提在现实情形中几乎当然非真——肯定存在不喜欢红葡萄酒的法国人。然而这并不是重点。使这个推论成立的是存在于前提和结论之间的一种恰当关系,即前提为真则结论也必然为真。前提实际上是否为真则是另外一回事,它并不影响推论的演绎性质。

　　并非所有的推论都是演绎的。请看下面的例子:

盒子里前五个鸡蛋发臭了
所有鸡蛋上标明的保质日期都相同

因此,第六个鸡蛋也将是发臭的

　　这看起来似乎是一个非常合理的推理。但它却不是演绎性的,因为前提并不必然导致结论。即使前五个鸡蛋确实发臭了,并且即使所有鸡蛋上标明的保质日期相同,也不能保证第六个鸡蛋一样发臭。第六个鸡蛋完好无损的情况是很有可能的。换言之,这个推论的前提为真而结论为假,这在逻辑上是可能的,所以这个推论不是演绎的。它被称为归纳推论。在归纳推论或者说归纳推理中，我们是从关于某对象已被检验的前提推论到关于该对象的未被检验的结论——本例中这个对象是鸡蛋。

演绎推理是一种比归纳推理更可靠的推理方式。进行演绎推理时，我们可以保证从真前提出发就会得出一个真结论。然而，这种情况却不适用于归纳推理。归纳推理很有可能使我们从真前提推出一个假结论。尽管存在这种缺点，我们却似乎一直都在依赖归纳推理，甚至很少对它进行思考。例如，当你早上打开电脑的时候，你相信它不会在你面前爆炸。为什么呢？因为你每天早上都会打开电脑，它至今从来没有在你面前爆炸过。但是，从"迄今为止，我的电脑在打开时都不曾爆炸过"到"我的电脑在此时打开时将不会爆炸"的推论是归纳的，而不是演绎的。这个推论的前提并不必然得出结论。你的电脑此时爆炸在逻辑上是可能的，即使这在以前从来没有发生过。

在日常生活中，其他归纳推理的例子随处可见。当你逆时针转动方向盘的时候，你认为汽车将会向左而不是向右拐。驾车上路时，你就把生命作为赌注压在这一假定上。是什么使你如此肯定它是正确的？如果有人要求你证明这一确信，你将如何回答？除非你是一个机修工，你有可能会回答："在过去每一次我逆时针转动方向盘的时候，汽车都是向左拐的。因此，当我这一次逆时针转动方向盘时也会发生同样的情况。"这同样是归纳推论，而不是演绎推论。归纳性推理似乎是我们日常生活中不可或缺的一部分。

科学家们也运用归纳推理吗？答案似乎是肯定的。来看一种被称为唐氏综合征的遗传学疾病。遗传学家告诉我们唐氏综合征患者具有一条多余的染色体——他们拥有 47 条而不是常人的 46 条（参见图 5）。他们是如何发现的呢？答案当然是，他们测试了大量的唐氏综合征患者并且发现每一位患者都有一条多余的染色体。于是他们便归纳地推出这一结论，即所有的唐氏综合征患者，包括尚未进行检验的，都有一条多余的染色体。很容易看出这个推论是归纳的。研究样本中的唐氏综合征患者有47条染色体的事实，并不能证明所有的唐氏综合征患者都是如此。尽管不太可能出现这样的情况，但是一个非典型样本的存在也是有可能的。

这种例子绝不仅仅只有一个。事实上，无论何时从有限的资料数据获得一个更普遍的结论，科学家们都要运用归纳推理，这是他们一直使用的方法。以牛顿的万有引力原理为例，如上一章所述，该定律讲的是宇宙中的每一个物体都会对任一其他物体产生引力作用。很显然，牛顿并没有通过检验宇宙中的每一物体来得出这一定律——他不可能这样做。其实，他首先发现行星和太阳以及地球表面附近各种运动的物体适用这个定律。从这些数据中，他推论出定律对于所有的物体都适用。这一推论显然也是归纳性的：牛顿定律适用于某些物体的事实并不能保证它适用于所有物体。

归纳在科学中的核心作用有时候是被我们的说话方式弄得含糊不清了。例如，你也许看到报上说科学家已经"通过实验证明"基因改良的玉米对人体是安全的。这里的意思是科学家已经对于大量的人测试了这一种玉米，没有一个人产生任何不良反应。但是，严格地说这并没有**证明**这种玉米是安全的，即没有像数学家证明毕达哥拉斯定理那样。因为，从"这种玉米对于被检验过的人没有任何坏处"到"这种玉米对于任何人都没有坏处"的推论是归纳的，而不是演绎的。这份报纸本来应该如实地说，科学家已经发现了特别有力的**证据**表明这种玉米对人

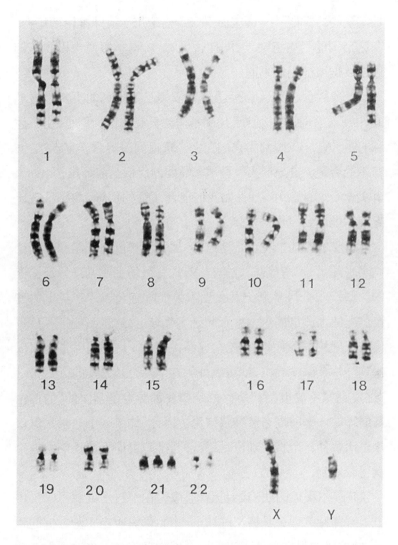

图 5 唐氏综合征患者整套染色体(或者说染色体组型)示意图。不像多数正常人的 21 号染色体具有两条复制体，唐氏综合征患者的 21 号染色体有 3 条复制体，因此他们总共有 47 条染色体。

是安全的。"证明"一词应该仅仅严格地用于演绎推论的场合。在这个词的严格意义上,即使曾经有过,科学假说也极少能够通过数据被证明是真的。

大多数哲学家认为科学过分依赖归纳推理的事实是显然的,由于过于明显,以至于几乎不需要再有辩论。但是,引人注意的是,这一点遭到了我们在上一章提到过的哲学家卡尔·波普尔的否定。波普尔认为科学家需要的仅仅是演绎推论。如果事实真是如此就好了,正如我们已经了解的那样,演绎推理比起归纳推理要可靠得多。

波普尔的基本观点是这样的:尽管不可能证明某科学理论确实来源于一个有限的数据样本,却有可能证明某理论是错误的。假设一个科学家一直在思考关于所有金属片都导电的理论。即使她测试的每一片金属确实都导电,这也不能证明该理论是正确的,其原因我们上文已经说清楚了。但是哪怕她仅仅找到一片金属不导电,就可以证明这个理论是错误的。因为,从"这片金属不导电"到"所有的金属片都导电是错误的"的推论是演绎的——前提必然导致结论。因此,如果一个科学家仅仅热衷于解释一个特定理论是错误的,她有可能不使用归纳推论就可以做到。

波普尔观点的缺陷显而易见,原因在于科学家并不仅仅热衷于解释特定理论是错误的。当一个科学家收集实验数据的时候,她的目的也许是为了表明一个特定理论——也许是与她针锋相对的理论——是错误的。但更有可能的是,她正致力于说服人们相信她自己的理论是正确的。为了达到此目的,她将不得不求助于归纳推理。所以,波普尔想表明科学可以不需要归纳是不会成功的。

休谟的问题

虽然归纳推理在逻辑上并非无懈可击,但它似乎是形成关于世界之信念的一种非常合理的方法。迄今为止太阳每天都升起的事实也许不能证明明天它会升起,但是这一事实是否的确给了我们很好的理由相信太阳明天会升起?如果你遇到某个人声称完全拿不准明天太阳是否会照常升起,你即使不说他神智不清,也一定会把他视为非常古怪的人。

然而,是什么证明了我们对于归纳的信任是正确的?我们该怎样说服拒绝归纳推理的人他们是错误的? 18 世纪苏格兰哲学家大卫·休谟(1711—1776)对这一问题给出了一个简单而又激进的答案。他认为,运用归纳的正当性不可能完全从理性上被证明。休谟承认,我们在日常生活和科学活动中时刻都在运用归纳方法,但是他主张这仅仅是一种与理性无关的动物性习惯。他认为,若要为归纳的运用提供充分的理由,我们不可能办得到。

休谟如何推出这一令人惊讶的结论?他首先提出,无论我们何时进行归纳推论,似乎都要预设他所称的"自然的齐一性"。为了弄清休谟在这里的意思究竟是什么,我们再回顾一下上一节关于归纳推理的一些内容。我们考察了从"我的电脑至今没有爆炸过"到"我的电脑今天也不会爆炸";从"所有被检验的唐氏综合征患者都有一条多余的染色体"到"所有唐氏综合征患者都有一条多余的染色体";从"至今观察的所有物体都遵守牛顿的引力定律"到"所有的物体都遵守牛顿的引力定律"等等这些推论。对于这些案例中的每一情形,我们的推理似乎都依赖于一个假设,即我们未检验过的物体将在某些相关的方面与我们已经检验过的同类物体相似。这一假设正是休谟对于自然的齐一性的解释。

但是正如休谟所问,我们如何获知自然的齐一性假设实际上是正确的呢? 我们能否在某种程度上证明(严格意义上的证明)它的正确性呢?不,休谟说,我们不可能做到。因为我们很容易想到,宇宙并不是齐一的,并且宇宙每天都在任意地改变。在这样一个宇宙中,电脑有时也许会无缘无故地发生爆炸,水有时也许会在毫无征兆的情况下使我们中毒,台球在碰撞中也许会停止运动,等等。既然这样一个"非齐一"的宇宙是可能存在的,我们就不可能严格证明自然的齐一性的正确性。原因在于,如果我们可以证明其正确性,这个非齐一的宇宙在逻辑上就不可能存在了。

自然的齐一性虽不可证明,我们却有可能寄望于找到证明其正确性的经验证据。毕竟,迄今为止自然的齐一性一直保持其正确性, 这是否就的确给了我们很好的理由相信它是真的?休谟认为,这种观点回避了我们的问题! 因为它本身就是一个归纳推理,所以它本身就要依赖自然的齐一性的假设。一个从一开始就假定自然的齐一性的观点,显然不可能被用来证明自然的齐一性是正确的。换一种方式来说:一个确定的事实是至今为止自然在大体上是齐一的,但是我们不能引用这一事实去论证自然将持续齐一,因为它假定过去已经发生的情况能可靠地标示未来将会发生的情况——这正**是**自然的齐一性假设。如果我们试图依靠经验来论证自然的齐一性,我们就会陷入循环推理。

休谟观点的力量可以通过下述情况来理解,即设想你如何去说服本该相信而不相信归纳推理的人。你也许会说:"看,归纳推理至今都在发挥着很好的作用。通过运用归纳方法科学家已经分裂了原子、使人类登上月球、发明了计算机,等等。反之,那些不曾运用过归纳方法的人已经走向痛苦的死亡。他们吞下砒霜,认为它们能滋养身体,从高楼上跳下,认为可以凌空飞翔,等等(参见图6)。因此,运用归纳推理显然会让你受益匪浅。"但是,这当然无法说服怀疑者。因为,声称归纳值得信赖是因为它迄今为止都发挥着很好的作用,这本身就是以一种归纳的方式在进行推理。对于尚不信任归纳方法的人来说,这种观点是没有说服力的。此即休谟的基本观点。

这就是问题所在。休谟指出,我们的归纳推论建立在自然的齐一性假设之上。但是我们无法证明自然齐一性是正确的,并且我们只有回避这一问题才能为它的正确性提供经验性证据。所以,我们的归纳推论依据的是一种关于世界的假设,对于该假设我们没有很好的根据。休谟断定,我们对于归纳的信心只是盲目的确信——无论如何它无法在理性上得到辩护。

这种引人兴趣的观点已经在科学哲学领域产生了巨大的影响,并且这种影响力今天仍在持续。(波普尔的一个失败的论证,即科学家仅仅需要运用演绎推论方法,就源于他相信休谟已经表明归纳推理完全是非理性的。)休谟观点的影响并不难理解。在通常情况下,我们认为科学正是理性探究的范式。对科学家们所说的关于世界的一切,我们深信不疑。每一次坐飞机旅行,我们都把自己的生命放在设计飞机的科学家的手上。但是科学却依赖着归纳,休谟的观点似乎表明归纳不可能被理性地辩护。如果休谟正确,建立科学的基础看起来就不如我们所希望的那样坚固。这种使人困惑的情形被称为休谟归纳问题。

哲学家们已经用了差不多数十种方法来回应休谟的问题;

科学推理

图 6 对于那些不相信归纳方法的人所发生的情况。

在今天,这一问题仍然是研究的热点领域。有些人认为,问题的关键在概率这个概念上。这种提法似乎非常合理。因为人们很自然地就可以想到,尽管一项归纳推论的前提不保证结论正确,但它们确实使结论非常有可能成立。同样,即使科学知识并不是确定的,但它为真的概率仍然很大。但是,对于休谟问题的这种回应又产生了它自身的难题,并且这种回应绝不会被广泛接受;我们将在适当的时候再讨论这一点。

另一个常见的回应是:承认归纳不可能在理性上得到辩护,但是主张这一点事实上并不成问题。人们是如何为这种主张作辩护的呢?一些哲学家已经指出,归纳对于我们思考和推理是如此重要,以至于它并不是那种正当性可以被证明的东西。彼得·斯特劳森,当代一位颇有影响的哲学家,为辩护这种观点作出了以下类比:如果有人担心一个特定的行为是否合法,他们可以查阅法律书籍并把这一行为同法律书上所写的内容作比较。但是若有人担心法律本身是否合法,这的确就是一种很奇怪的担心了。因为法律是判断其他事情的合法性的标准,探究这种标准本身是否合法几乎是没有意义的。斯特劳森认为,同样的情况也适用于归纳。归纳是一种我们用来决定关于世界的断言是否正确的标准。例如,我们运用归纳来判断一个制药公司关于它的新药物利润惊人的看法是否正确。因此,问归纳本身是否正当是无意义的。

斯特劳森真的成功解决了休谟问题吗?一些哲学家认为是的,另一些哲学家认为不是。但是大多数人都同意,为归纳作出一个令人满意的辩护非常困难。(弗兰克·拉姆齐,一位来自 20 世纪 20 年代剑桥大学的哲学家,认为试图为归纳寻求辩护就等于试图"水中捞月"。)这个问题是否应该使我们担心或者动摇我们对科学的信念,是一个你自己应该深思熟虑的难题。

最佳说明的推理

我们至今为止考察过的归纳推论事实上都拥有同样的结构。在每一个例子中,推论的前提都具有这样的形式:"迄今为止所有验证过的 x 都是 y",结论具有的形式是"下一个将要验证的 x 也会是 y",或者有时说"所有的 x 都是 y"。换言之,这些推论使我们从某种条件下经过验证的情形推出某种条件下未加验证的情形。

正如我们所看到的,这样的推论被广泛应用在日常生活和科学活动中。然而,还存在不符合这种简单模型的另外一种普通非演绎性推论。请看下面的例子:

食品柜里的干酪不见了,仅留下一些干酪碎屑
昨天晚上听到了来自食品柜的刮擦声音

所以,干酪是被老鼠吃了

显然这一推理是非演绎性的:前提并不必然导致结论。干酪有可能是被女仆偷了,她巧妙地留下一些碎屑以使这看起来像是老鼠的杰作(参见图 7)。刮擦声响可以由许多方式造成——也许是由于水壶加热过头。尽管如此,这个推论却显然是一个合理的推论。假设老鼠吃掉了干酪似乎比其他各种解释都更为合理。毕竟,女仆通常是不会偷干酪的,现代的水壶一般也不会加热过头。而老鼠通常却一有机会就会偷吃干酪,并且的确会制造些刮擦的声响。因此,虽然我们不能确定猜想老鼠

图 7 老鼠假说与女仆假说二者都可以作为失踪干酪的解释。

作案是对的,但总体来讲这个假说看起来相当合理:它是对已
知事实最好的解释方式。

　　因为很显然的原因,这种类型的推理被称为"最佳说明的
推理"(inference to the best explanation),或者缩略为 IBE。围绕
着 IBE 和归纳推论之间的关系产生了某些术语混乱。有些哲学
家把 IBE 归为归纳推论的一种;实际上,他们使用"归纳推论"
一词指的是"任何一种非演绎的推论"。另外一些哲学家把 IBE

和归纳推论对立看待,正如我们在前文所做的。按照这种划分方式,"归纳推论"专指从某种既定的已检验的情形推出某种未检验的情形,即我们在之前考察过的那类情形;IBE 和归纳推论因此是两种不同类型的非演绎推论。只要我们坚持这一点,在选择术语时就不会有什么为难之处。

科学家们频繁地使用 IBE。例如,达尔文通过唤起对生物世界各种现象的关注来论证他的进化论理论,认为如果假设现存的物种都是孤立地被创造出来,生物世界的现象就很难得到解释,但是如果像他理论中所说的,现存物种是从共同的祖先演化而来,就很容易说得通。例如,马和斑马的腿二者在解剖学上具有紧密的相似性。如果上帝分别创造了马和斑马,我们如何解释上述情况呢?可以想见,只要愿意,上帝本可以把它们的腿做得大为不同。但是,如果马和斑马二者是由上一级共同祖先演化而来的,这就为它们的解剖学相似性提供了一种显明的解释。达尔文认为,他的理论对于这种现象以及其他许多种现象的解释力,为其正确性提供了强而有力的证据。

另一个 IBE 的例子是爱因斯坦对于布朗运动的杰出贡献。布朗运动指的是悬浮在液体或气体中的微小粒子所作的无规则、曲折的运动。它是由苏格兰植物学家罗伯特·布朗(1713—1858[①])在 1827 年观察水中漂浮的花粉粒时发现的。19 世纪出现了许多种试图解释布朗运动的理论。一种理论把运动归因于粒子间的电荷吸引力,另一种理论将其归于来自外在环境的扰动作用,还有一种理论则归因于液体内的对流作用。正确的解释建立在物质动力学说之上,该理论认为液体和气体都是由运

---

① 原文如此。实际应为 1773—1858。——编注

动着的原子和分子组成的。悬浮的微粒与周围的分子发生碰撞，导致了由布朗第一个观察到的无规律的、任意的运动。这一理论是在 19 世纪后期首先被提出的，但并没有得到广泛接受，在一定程度上是因为许多科学家并不相信原子和分子是真实的物理实体。但是在 1905 年，爱因斯坦对于布朗运动进行了独创性的数学分析，作出了许多精确的、定量的预测，这些预测都被后来的实验所证实。在爱因斯坦的研究之后，分子运动论很快被认为是对布朗运动提供了一种远远优于其他理论的解释，对于原子和分子是否存在的质疑很快就平息了。

一个有趣的问题是，IBE 和普通的归纳哪一种是更基本的推论模式。哲学家吉尔伯特·哈曼认为 IBE 更基本。按照这种观点，无论何时作出诸如"至今所有已检验的金属片都导电，所以所有的金属片都导电"这样的普通归纳推论，我们暗中都在诉诸解释性的观点。我们假设的是，对于样本中的金属片为何导电的正确解释，无论它是什么，都必然推出所有的金属片导电；这就是我们进行归纳推理的原因。但是如果我们相信，样本中的金属片之所以导电是因为（例如）一个实验人员对其进行了处理，我们就不会推出所有的金属片都导电。这一观点的支持者并不是认为 IBE 和一般的归纳之间没有任何差别——差别显然是有的，而是认为普通的归纳最终要依赖于 IBE。

其他的哲学家则认为这正好颠倒了事实：他们认为，IBE 本身依附在普通的归纳之上。为了找到支持这种观点的理由，让我们回顾一下上文的食品库干酪的例子。我们为什么认为老鼠假说是一种比女仆假说更好的解释呢？大概是因为，我们知道女仆通常是不偷吃干酪的，而老鼠却不然。但是，这是我们从普通的归纳推理中得到的知识，建立在我们对于老鼠和女仆先前行为观察的基础之上。所以按照这种观点,若想决定在一组竞争性的假说中哪一个是对事件的最佳解释,我们总是诉诸从普通的归纳中获得的知识。因此,认为 IBE 是一种更基本推论模式的观点是不正确的。

　　无论偏向于上述对立观点中的哪一种,有一点显然要引起更多的关注。若想使用 IBE,我们需要某种方法来确定竞争性假说中哪一个提供了对事实的最佳解释。通过什么标准来确定这一点呢? 一种常见的答案是,最佳解释指的是最简单或者原因最少的解释。再回顾一下食品柜干酪的例子。其中有两个事实需要解释:丢失的干酪和刮擦的声响。老鼠假说仅仅假定了一个原因——老鼠——来解释两个事实。女仆假说必须假定两个因素——一个不诚实的女仆和一个过度加热的水壶——作为条件来解释同样的事实。所以老鼠假说假设的原因更少,因此更好。在达尔文的例子中也是如此。达尔文的理论能够解释关于生物世界非常广泛的事实,并不止是物种之间解剖学上的相似性。正如达尔文了解的那样,这些事实中的每一个都可以通过其他的方式得到解释。但是进化论能一揽子解释所有的事实——这就是使它成为最佳解释的原因。

　　简单性或简洁性是一个好的解释的标志——这一观点相当有吸引力,并且对于充实 IBE 观点确有帮助。但如果科学家运用简单性作为进行推论的指导,就会产生一个问题。我们如何知道宇宙是简单而不是复杂的呢? 偏爱以最少的原因来解释事实的理论看似的确有理。但是,对比不如它简单的理论,是否存在客观的理由支持它更有可能正确呢?科学哲学家在这个难题上并没有达成一致意见。

概率与归纳

概率的概念在哲学上令人困惑。部分的困惑在于,"概率"这个词似乎具有不止一种含义。如果听说英国妇女寿命达到100岁的概率是十个中有一个,你会把该信息理解为有十分之一的英国妇女寿命达到了100岁。同样,如果听说男性吸烟者患肺癌的概率是四个中有一个,你会认为这指的是有四分之一的男性吸烟者患肺癌。这被称为概率的频率意义上的解释:它把概率等同于比例,或者说频率。但是,如果你看到在火星上发现生命的概率是千分之一,你会如何理解呢? 这意味着太阳系中每一千个行星中就有一个含有生命吗?显然不是。首先,太阳系中仅仅存在九个行星。因此,概率在这里必定有另一种所指。

对于"火星上存在生命的概率是千分之一"的一种解释是,如此陈述的人只是在表达他们自己的主观想法——告诉我们对于火星上存在生命他们认为有多大的可能性。这是概率的主观意义上的解释。它把概率作为衡量我们个人信念强弱的一种尺度。显然,我们对自己所持的某些信念比其他的信念要更为坚定。我非常有信心巴西队会夺得世界杯,也相当相信耶稣基督的存在,但不怎么相信全球环境灾难可以被避免。这一点可以通过以下的陈述来表达:我对"巴西队会夺得世界杯"这一说法赋予很高的概率,对"耶稣基督的存在"赋予较高的概率,对"地球环境灾难可以被避免"赋予较低的概率。当然,要给这些陈述的信念强度标出精确的数值是很难的,但主观式解释的支持者认为这仅仅是实践上的不足。他们认为,在原则上,我们应该能够对每一种陈述赋予一个精确的用数字表示的概率,来反映我们相信或不相信这些陈述的强度有多大。

概率的主观解释暗示了不存在关于概率的客观的、独立于人之信念的事实。我说火星上存在生命的概率很高，而你说这个概率很低，我们之间没有谁对谁错——我们都仅仅是在表达我们对相关陈述有多强的信念。当然，关于火星上是否有生命，客观的事实是存在的；但是按照主观解释，在火星上有生命的可能性有多大这一点上则不会有客观的事实存在。

概率的逻辑学解释拒绝接受这种立场。它认为诸如"火星上有生命的概率很高"这一陈述存在客观上的对错之别，它与一组特定的证据相关。按照这种观点，一个陈述的概率即支持该陈述的证据强弱的尺度。逻辑学解释的支持者认为，在用语言所作的任何两个陈述中，我们在原则上可以得出其中一个陈述的概率，而把另一个陈述作为证据。例如，已知现在地球变暖的速度，我们想要得出在一万年之内出现冰川期的概率。主观式解释认为，关于这种概率的客观事实并不存在。但逻辑解释坚持认为存在：现在地球变暖的速度对于一万年内出现冰川期的陈述赋予了一个确定的可用数字表达的概率，比如说是 0.9。0.9 的概率显然很高——最大值是 1，所以在给定地球变暖的证据下，"一万年内出现冰川期的概率很高"这一陈述在客观上就是正确的。

如果已经学习了概率论或统计学，你也许会对上述概率有不同解释的说法感到困惑。这些解释和你所学的如何相关？答案是：概率的数学研究本身并没有告诉我们概率的含义，这一含义正是我们在上文所一直探究的。事实上，大多数的统计学家会倾向于概率的频率式解释，但是，关于如何解释概率的问题，正如大多数哲学问题一样，不可能以数学的方式得到解决。不论采用哪一种解释，计算概率的数学公式都是一样的。

科学哲学家对概率感兴趣主要出于两个原因。一是在许多科学分支,特别是物理学和生物学中,定律和理论都是运用概率这个概念得出的。例如,以著名的孟德尔遗传学理论为例,它研究的是有性繁殖的种群中基因的代际传递问题。孟德尔遗传学最重要的原理之一是,有机体中的每一个基因使自身成为该有机体的配子(精子或卵子)的机会都是 50%。因此,你母亲身上的任何基因将有 50%的机会也存在于你的体内,你父亲身上的基因也同样如此。运用这一原理与其他的原理,遗传学家能够提供详细的解释,即为什么某些显著的特性(例如眼睛的颜色)以现有方式在家族的代际之中分配。在此,"机会"就是概率的另一种表达, 孟德尔遗传学原理显然切实运用了概率的概念。还可以给出许多其他的例子,这些例子都是通过概率来表达科学定律和原理的。理解这些定律和原理的需要,是对概率进行哲学研究的一个重要动力。

科学哲学家对概率这个概念感兴趣的第二个原因,是希望可以借助它阐明归纳推论,特别是休谟问题;这一点就是我们在此所关注的。休谟问题的根源是这样一个事实:一个归纳推论的前提条件并不保证其结论为真。然而人们很容易声称,一个典型归纳推论的前提条件确实为结论赋予了很高的可能性。尽管至今为止所有被检验的物体都遵循牛顿万有引力原理这一事实,并不能证明所有的物体都是这样,但该事实肯定就使得所有物体都遵循牛顿万有引力原理很有可能吗?休谟问题真的很容易得到解答吗?

事情并非如此简单。我们必须追问,对休谟的回应采取的是哪一种概率解释方式。按照频率式解释,说极有可能所有物体都遵循牛顿定律,就是指所有的物体中有很大比例遵循这一定律。但是除非运用归纳法,我们无法知道这一点!我们仅仅验证了宇宙所有物体中很小的一部分。因此,休谟问题仍然存在。看待这一点的另一方式是这样的:我们先看从"所有已检验的物体都遵循牛顿定律"到"所有物体都遵循牛顿定律"这一推论。为了回应休谟的担心,即这一推论的前提不保证结论为真,我们设想即便如此它却使结论很可能成立。但是,从"所有已检验的物体都遵循牛顿定律"到"所有物体都遵循牛顿定律很可能成立"仍然是归纳推论,鉴于后者意指"所有物体中有很大比例遵循牛顿定律",正如频率式解释的情形。所以,若采用概率的频率式解释,诉诸概率这个概念就并不能解决休谟的问题。因为这样的话,关于概率的知识本身就得依靠归纳。

概率的主观式解释对于休谟问题同样无能为力,尽管原因有所不同。假设约翰认为太阳明天将会升起而杰克认为它不会升起。两人都接受在过去太阳每天都升起的证据。在直觉上,我们会说约翰是理性的而杰克则不是,因为证据使约翰所相信的更为可能。但是如果概率仅是一个主观观念的问题,我们就不能这样说。我们所有能说的只是,约翰对于"太阳明天升起"赋予了一个很高的概率而杰克没有。如果不存在关于概率的客观事实,我们就不能说归纳推论的结论在客观上是可能的。所以我们就无法解释为什么像杰克那样拒绝使用归纳方法的人是不理性的。然而,休谟问题正需要这样的一个解释。

概率的逻辑学解释更有希望在休谟问题上作出令人满意的回应。鉴于太阳在过去的每一天都升起了,我们假设有一个关于太阳明天将会升起这一概率的客观事实。假设这一概率非常高。由此,我们就能解释为什么约翰是理性的而杰克不是。因为,约翰和杰克两人都接受了太阳过去天天升起的证据,但是杰克没有意识到这一证据使得太阳明天升起很可能成立,而约翰却意识到了这一点。正如逻辑学解释所建议的,把一个陈述的概率看做是支持它的证据的衡量尺度,这与我们的直观感觉——归纳推论的前提条件可以使结论很可能成立,即便不能保证其正确性——巧妙地吻合。

因此,那些试图通过概率概念来解决休谟问题的哲学家倾向于支持逻辑学解释就不足为奇了。(其中之一就是著名经济学家约翰·梅纳德·凯恩斯,他的早期兴趣在于逻辑学和哲学。)不幸的是,今天的大多数人认为概率的逻辑解释面临着非常严重的、可能无法克服的难题。这是因为,在任何细节上完成概率的逻辑学解释的尝试都碰到了一堆问题,既有数学上的也有哲学上的。结果是,今天许多哲学家倾向于彻底拒斥逻辑学解释的基本假设——在给出另一个客观事实的情况下,存在关于一个陈述之概率的客观事实。拒斥这种假设自然就导向了概率的主观解释,然而正如我们已经了解的,主观解释在休谟问题上提供令人满意的回应希望渺茫。

即使休谟问题如看起来那样无望最终解决,关于这一问题的思考仍然有其价值。对于归纳问题的思考引导我们进入了一个有趣的问题之域,这些问题关乎科学推理的结构、理性的本质、人类依赖科学的适当限度、概率的解释,等等。与大多数哲学问题一样,这些问题可能没有终极答案,但是在探究它们的同时我们也对科学知识的本质和界限了解了很多。

第三章
科学中的解释

　　科学最重要的目的之一就是试图解释我们周围世界中所发生的一切。有时候,我们会出于实际的目的寻求解释。例如,我们也许想知道为什么臭氧层损耗的速度这么快,从而试着对它采取一些措施。在其他情况下,我们寻求科学解释仅仅是出于猎奇心理——我们想对这个世界了解地更多。在历史上,对科学解释的追求是由这两个目标共同推进的。

　　在提供解释这一目的上,现代科学常常能够成功。例如,化学家能够解释为什么钠在燃烧时变黄。天文学家能够解释为什么日食会出现。经济学家可以解释为什么日元在 20 世纪 80 年代贬值。遗传学家可以解释为什么男性秃头易于在家族内部遗传。神经生理学家可以解释为什么极度缺氧会导致大脑损伤。也许你还能想到许多其他成功的科学解释的例子。

　　但是,科学解释确切地说**是**什么呢? 说一个现象能够被科学进行"解释"究竟是什么意思?这是一个自亚里士多德开始就引起哲学家思虑的问题,但是我们将以美国哲学家卡尔·亨普尔在 20 世纪 50 年代对科学解释作出的著名阐释作为论述的起点。亨普尔的阐释被称为解释的**覆盖律**模型,其名称的由来在下文会有交待。

亨普尔的覆盖律解释模型

覆盖律模型背后的基本思想是直截了当的。亨普尔指出，科学解释通常是在回应被他称为"寻求解释的原因类问题"时给出的。这些问题包括诸如"为什么地球不是完全圆球形的？"、"为什么女人的寿命比男人长？"等——它们都寻求解释。给出科学解释因此就成了对寻求解释的原因类问题提供满意的答案。若能确定这个答案必须具有的本质特征，我们就会知道科学解释指的是什么。

亨普尔认为，科学解释的典型逻辑结构和论证是一样的，即由一系列前提得出一个结论。结论断言待解释的现象实际发生了，前提则告诉我们这个结论为什么正确。这样，我们可以设想有人询问为什么糖在水中会溶解。这就是一个寻求解释的原因类问题。要回答它，亨普尔认为，我们必须构建一个论证，其结论是"糖在水中溶解"，前提则告诉我们为什么这个结论是正确的。如此一来，为科学解释提供描述的任务就变成了准确地刻画一组前提和一个结论之间必定具有的关系，从而把前者看做对后者的解释。这就是亨普尔为自己设定的问题。

亨普尔对于这一问题的回答分三个层次。首先，前提应该保证推出结论，即论证应该是演绎推理。第二，前提应该全部为真。第三，前提应该至少包含一个普适定律。普适定律指的是诸如"所有的金属都导电"、"一个物体的加速度与它的质量成反比变化"、"所有的植物都含有叶绿素"等等；它们与诸如"这一片金属导电"、"我书桌上的植物含有叶绿素"等特殊事实相对。普适定律有时也被称为"自然律"。亨普尔承认，科学解释也会像求助于普适定律那样求助于特定事实，但是他认为至少一个普适定律总是必需的。因此按照亨普尔的观念,解释一个现象就是去表明它的出现可以从一个普适定律演绎地推出,也许还要补充其他的定律和／或特定事实,它们都必须是正确的。

为了解释这一观点,假设我正尝试解释为什么书桌上的植物死掉了。我可能给出如下解释:我学习的地方光线太暗,阳光无法照射到植物上；而阳光是植物进行光合作用所必需的；并且没有光合作用,植物就不能制造它存活所必需的碳水化合物,因此就会死掉。这一解释完全符合亨普尔的解释模型,它对植物死亡的解释是通过从两个正确的定律(阳光是光合作用所必需的以及光合作用是植物存活所必需的)和一个特定事实(植物没有受到任何阳光的照射)进行演绎而得出的。由于这两个法则和特定事实的正确性,植物的死亡就**不得不**发生了；于是前者构成了对后者的一个很好的解释。

亨普尔的解释模型可以用如下示意图来描述:

普适定律

特定事实

=〉

待解释的现象

待解释的现象被称为**被解释项**,用做解释的普适定律和特定事实被称为**解释项**。被解释项本身或者是一个特定事实,或者是一个普适定律。在上面的例子中,它是一个特定事实——我的植物的死亡。但是有时候,我们想要解释的对象具有普遍性。例如,我们会希望解释在太阳下暴晒导致皮肤癌的原因。这是一个普适定律,而不是特定事实。为了解释它,我们需要从更

科学哲学

加基础的法则——大约是光线对皮肤细胞影响的法则——出发进行演绎，并结合关于太阳辐射能量的特定事实。因此，不管被解释项(即我们试图解释的事物)是特定的还是普通的，科学解释的结构在本质上是一样的。

很容易看出为什么亨普尔的模型被称为覆盖律解释模型：按照这一模型，解释的本质就是表明待解释的现象是被某个自然普适定律所"覆盖"的。这种观点确实有吸引人之处。因为，表明一个现象是某个普适定律的结果确实在某种意义上祛除了它的神秘性——使它更易于理解。事实上，科学解释的确经常符合亨普尔所描述的形式。例如，牛顿解释了为什么行星在椭圆轨道上围绕太阳旋转的现象，表明这可以由他的万有引力原理以及一些次要的附加假设演绎地推导出来。牛顿的解释完全符合亨普尔的解释模型：这里对现象的解释方式是，在自然律以及一些附加事实面前，一个现象不得不如此。牛顿之后，为什么行星轨道是椭圆形的这个问题就不再神秘了。

亨普尔知道，并不是所有的科学解释都完全符合他的模型。例如，如果你问人为何雅典总是沉浸在烟雾之中，他们可能会说："因为汽车的尾气污染。"这是一个完全可以接受的科学解释，尽管它并没有涉及任何定律。然而亨普尔会说，如果该解释被详细地表达出来，定律就会被涉及。可能存在一个类似这样的定律："如果一氧化碳以足够大的密度被排放到地球大气层，烟雾云层就会形成。"对于雅典为什么沐浴在烟雾中的充分解释将会援引这一定律，以及如下事实：汽车尾气中含有一氧化碳；雅典有很多汽车。在实践中，我们不会作出如此详细的解释，除非是学究气十足的人。但是一旦我们想解释清楚，它就会同覆盖律形式相当吻合。

从他关于解释与预测之联系的的模型中,亨普尔得出了一个有趣的哲学结论。他认为,解释和预测是同一个硬币的两面。无论何时对一个现象进行覆盖律解释,我们本来都可以利用所引用的规律和特定事实预测出该现象的发生,即使我们并不知晓。为了解释这一点,我们再看一下牛顿对于行星轨道为什么是椭圆的解释。这一事实早在牛顿用他的引力理论进行解释之前就为人所知——发现者是开普勒。但是即使不为人知,牛顿本来也可以通过引力理论预测出来,因为他的理论与次要的附加假设相结合必然推出行星的轨道是椭圆的。亨普尔是通过以下说法来表达这一点的:每一个科学解释都潜在地是一个预测——它可以用来预测相关现象,即使该现象还没有被了解。亨普尔认为反过来说也是正确的:每一个可靠的预测都潜在地是一种解释。为了说明这一点,假设科学家根据山区大猩猩生活环境遭破坏的信息,预测它们将会在 2010 年前灭绝。假设这一预测被证明是正确的。按照亨普尔的观点,他们在大猩猩灭绝之前用来进行预测的信息将可以在灭绝发生之后被用以解释同一事实。解释和预测在结构上是对称的。

尽管覆盖律模型很好地说明了许多现实的科学解释的结构,它仍然面临着许多棘手的反例。这些反例有两类。一方面,有一些真正科学解释的情形并不符合覆盖律模型,即使近似符合也算不上。这些情形表明亨普尔的模型太严格了——它把一些**真正的**科学解释排除在外了。另一方面,有一些情形**确实**符合覆盖律模型,但是直观上并不算真正的科学解释。这些情形又表明亨普尔的模型太随意了——它纳入了本该被排除在外的情况。我们将把焦点集中在第二类反例上。

对称问题

假设你正躺在阳光明媚的沙滩上,注意到一根旗杆在沙地上投射了一个 20 米长的影子(参见图 8)。

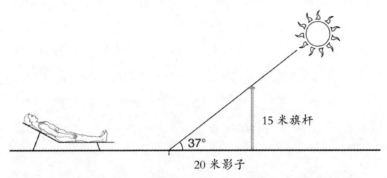

15 米旗杆

37°

20 米影子

图 8　当太阳在头顶 37° 仰角的位置时,一根 15 米长的旗杆在沙滩上投射出一个 20 米长的影子。

有人要求你解释为什么影子是 20 米长。这是一个寻求解释的原因类问题。一个合理的答案也许是这样的:"来自太阳的光线射到了旗杆上,旗杆整整 15 米高。太阳的仰角是 37°。由于光线照射路径是直线式的,简单的三角计算($\tan 37° = 15/20$)表明旗杆会投下 20 米长的影子。"

这看起来像是一个非常好的科学解释。通过改写使之与亨普尔的格式相一致,我们可以看出它是符合覆盖律模型的:

普适定律　　　　　光沿直线传播

　　　　　　　　　三角运算法则

特定事实　　　　　太阳的仰角是 37°

　　　　　　　　　旗杆 15 米高

43

=>

待解释的现象　　　影子 20 米长

　　旗杆的高度和太阳的仰角,连同光走直线的光学定律和三角运算法则一起,演绎推导出影子的长度。由于这些定律法则是正确的,并且由于旗杆的确是 15 米高,这一解释就精确地满足了亨普尔模型的要求。到现在为止,一切都很顺利。问题产生于下面的情况:假设我们将被解释项——影子 20 米长——换成旗杆 15 米高这一特定事实。结果是这样的:

普适定律　　　　　光沿直线传播
　　　　　　　　　三角运算法则

特定事实　　　　　太阳的仰角是 37°
　　　　　　　　　影子 20 米长

=>

待解释的现象　　　旗杆 15 米高

　　这一"解释"显然也符合覆盖律模型。旗杆投射影子的长度和太阳的仰角,连同光走直线的光学定律和三角运算法则一起,演绎推导出旗杆的高度。但是,若把这看做是对于旗杆为什么是 15 米高的**解释**,似乎非常怪异。旗杆为何是 15 米高的真正解释,推测起来应该是木匠故意地把它做成这样——它和它投射的影子长度毫无关系。因此,亨普尔的模型太不严格:它把显然不是科学解释的情形也看做科学解释。
　　旗杆例子的一般寓意是,解释的概念显示了一种重要的不对称。在给定相关定律法则和附加事实的情况下,旗杆的高度为影子的长度提供了解释,但是并不存在反之亦然的情况。一般来说,在给定相关定律法则和附加事实的情况下,如果 x 为 y 提供了解释,则在给定同样定律法则和事实的情况下,y 为 x 提供解释将不会是正确的。这有时也被说成:解释是一种不对称关系。亨普尔的覆盖律解释模型没有考虑这种不对称问题。因为,正如我们可以在给定定律和附加事实时,由旗杆的高度推出影子的长度,我们也可以由影子的长度推出旗杆的高度。换言之,覆盖律模型暗示着解释应该是一种对称关系,但事实上解释具有不对称性。因此,亨普尔的模型没有完全弄清什么才是科学解释。

对于亨普尔的解释和预测是同一硬币之两面的理论,影子和旗杆的案例也可以提供一个反例。原因很显然。假设你不知道旗杆有多高。如果有人告诉你它现在投下的影子是 20 米长、太阳在头上方 37°的位置,在了解相关光学和三角运算定律法则的情况下,你将能够**预测**出旗杆的高度。但是正如我们刚刚看到的,这一信息显然并没有**解释**旗杆为什么是那个高度。所以,在这一例子中预测和解释分道扬镳了。为我们未知的事实提供预测的信息并不能在我们知道之后用于解释这同一个事实,这是亨普尔理论的吊诡之处。

不相关性问题

假设一个小孩在一家医院一个挤满孕妇的房间里。小孩注意到房间里有一个人——一个名叫约翰的男性——没有怀孕,就问医生为什么。医生回答说:"约翰在过去的几年中一直有规律地服用避孕药。有规律地服用避孕药的人永远不会怀孕。因此,约翰没有怀孕。"为了讨论的需要,我们假设医生说的话是正确的——约翰有精神病并且确实服用了避孕药,他认为避孕药对他有益。即使这样,医生给小孩的答复也显然没有什么益处。很显然,约翰不怀孕的正确解释在于他是一名男性,而男性是不可能怀孕的。

但是,医生给予小孩的解释完全符合覆盖律模型。医生是从服用避孕药的人不会怀孕这一普适定律以及约翰一直在吃避孕药这一特定事实演绎推导出待解释的现象——约翰没有怀孕。由于普适定律和特定事实两者都是正确的,并且由于它们的确能保证推出被解释项,按照覆盖律模型,医生对于约翰为什么没有怀孕给出了一个相当充分的解释。但是,事实上他当然没有给出。所以覆盖律模型又是过于宽泛了:它把直观上并非科学解释的解释也作为科学解释接纳了进来。

这里总体的原则是,关于一个现象的良好解释应该包含与现象的发生**相关**的信息。这就是医生给孩子的回答出错的地方。尽管医生告诉小孩的话完全正确,约翰一直在服用避孕药的事实却与他没有怀孕的现象毫不相关,因为即使没有服用避孕药他也不会怀孕。这就是医生的回答算不上一个好答复的原因。亨普尔的模型并没有考虑到我们的解释概念的这一关键特征。

解释和因果性

由于覆盖律模型遇到了如此多的问题,寻找理解科学解释的其他替代路径就很自然了。有些哲学家认为问题的关键在于因果性这一概念。这是一个相当吸引人的主张,因为在多数情况下,解释一个现象事实上就是在解释是什么导致了它的产生。例如，一位事故调查者正在试图解释一起飞机坠毁事故，他正在寻找的显然是坠毁的原因。的确，"飞机为什么坠落"和"什么是飞机坠毁的原因"这两个问题实际上意思相同。同样，如果一个生态学家正在试图解释热带雨林地区的生物多样性为何不如过去，他显然正在寻找生物多样性减少的原因。解释和因果性这两个概念之间的联系相当紧密。

受这一联系的影响，许多哲学家已经放弃了对解释的覆盖律阐释而转向基于因果性的阐释。尽管内容有所变化，但这些解释背后的基本思想都是：解释一个现象无非就在于指出是什么导致了它的产生。在某些情况下，覆盖律和因果解释方式之间的差别实际上并不太大，因为从一个普适定律演绎推导出一个现象的发生常常就是给出它的原因。例如，再回顾一下牛顿对于行星轨道为什么是椭圆的解释。我们看到，这种解释是符合覆盖律模型的——牛顿从他的引力定律，再加上一些附加事实，推导出了行星轨道的形状。然而牛顿的解释也是一种因果方式，因为椭圆形的行星轨道是由太阳和行星之间的引力作用导致的。

但是，覆盖律和因果解释并非完全等同——在某些情况下它们是有分歧的。事实上，许多哲学家之所以倾向于解释的因果性阐释，正因为相信它可以避免覆盖律模型面临的一些问题。回顾一下旗杆的问题。为什么直觉告诉我们，在给定定律的情况下，旗杆高度为影子的长度提供了解释，但是却不能反之亦然？可信的回答是，因为旗杆的高度是导致影子 20 米长的原因，而 20 米长的影子却不是导致旗杆 15 米高的原因。所以，与覆盖律模型不同，解释的因果性阐释方式在旗杆案例中给出了"恰当的"解答——它考虑到了我们的直觉，即不能通过指出旗杆投射影子的长度来解释旗杆的高度。

旗杆问题的一般结论是,覆盖律模型不能体现解释是一种不对称关系这个事实。而因果性显然也是一个不对称关系：x是y的原因,y却并不是x的原因。例如,如果电路短路导致了火灾,显然火灾不会是导致短路的原因。因此,提出解释的不对称性来源于因果关系的不对称性似乎相当合情合理。如果解释一个现象就是去说出导致它产生的原因,那么,由于因果性是不对称的,我们就该预料到解释也具有不对称性——事实正是如此。覆盖律模型之与旗杆问题相冲突,正在于它试图在因果性之外分析科学解释的概念。

避孕药的例子也同样如此。约翰服用避孕药并没有解释他为什么没有怀孕,因为避孕药并不是导致他不怀孕的原因。实际上,约翰的性别才是他不怀孕的原因。正缘于此,我们认为"约翰为什么没有怀孕？"的正确答案是"因为他是一个男人,男人是不可能怀孕的",而不是医生提供的答案。医生的答案满足了覆盖律模型,但是,由于没有正确指出我们希望解释的现象产生的原因,它不构成一个真正的解释。我们从避孕药的例子中得到的一般结论是,一个真正的科学解释必定包含与被解释项相关的信息。实际上也就是说,解释应该告诉我们被解释项产生的原因。基于因果性阐释的科学解释与不相关性问题并不冲突。

亨普尔没有考虑到因果性与解释之间的密切联系,这一点很容易遭到人们的批评,并且许多人已经提出了批评。在某些方面,这种批评有失公允。亨普尔继承了被称为**经验论**的哲学纲领,而经验论者在传统上非常怀疑因果性这个概念。经验论认为我们所有的知识都来源于经验。我们在上一章提到过的大卫·休谟就是一位重要的经验论者，他认为因果联系不可能被经验到。他由此宣称因果联系是不存在的——它们只是我们想象中的虚构之物！这是一个让人很难接受的结论。摔下玻璃花瓶使它们破碎确实是一个客观事实吗?休谟认为不是。他承认，大多数玻璃花瓶在摔落之后发生破碎是一个客观事实，但我们关于因果性的观念包含的内容要比这更多。它包括在摔下和破碎之间的一个因果联系的观念,即前者导致了后者。按照休谟的观点,在世界上是找不到这种联系的:一个花瓶被摔,随后它破碎了,这就是我们看到的全部。我们并没有在第一个事实和第二个事实之间经验到任何因果联系的存在。因果性因此是一个虚构之物。

　　大多数的经验论者并没有完全接受这一令人惊讶的结论。但是由于休谟的主张,他们已经倾向于把因果性看做一个需要谨慎对待的概念。因此对于一个经验论者来说,使用因果性概念来分析解释这一概念似乎有违常理。如果一个人的目标是像亨普尔那样去澄清科学解释的概念,那么使用本身就需要澄清的概念来进行分析就没有什么意义。对于经验论者来说,因果性的确需要在哲学上加以澄清。因此,覆盖律模型没有关注因果性并不仅仅是亨普尔的疏忽。最近几年,经验论的受欢迎度在某种程度上已经降低了。另外,许多哲学家已经得出结论,认为因果性概念尽管在哲学上存在问题,但对于我们了解世界仍然不可缺少。因此,对科学解释基于因果性的阐释方式似乎比它在亨普尔时代更为人所接受。

　　对于解释基于因果性所作的阐释确实很好地抓住了许多实际科学解释的结构,但是仅止于此吗？许多哲学家认为不是这样,理由在于某些科学解释似乎并不具有因果性。一类例子来自科学中所谓"理论上的同一"。理论同一把一个概念等同于通常位于不同科学领域的其他概念。"水是 H_2O"就是一个例子,"温度是分子的平均动能"也是。在这两个例子中,一个熟悉的日常概念与一个比较深奥的科学概念是等同或者同一的。通常,理论同一为我们提供了类似于科学解释的东西。当化学家发现水是 H_2O 的时候,他们也就解释了水是什么。同样,当物理学家发现一个物体的温度是它分子的平均动能的时候,他们也就解释了温度是什么。而这两个解释都不是因果性的。由 H_2O 组成并不**导致**一种物质成为水——它仅仅**是**水。拥有特定的分子平均动能并不是**导致**一个液体具有它本身的温度——它仅仅是拥有那样的温度。如果这样的例子被接受为合理的科学解释,它们就表明,基于因果性阐释的解释并不能代表全部解释。

科学能解释一切吗?

现代科学能够解释我们所居住世界的大量事实。但是也有许多事实还没有得到科学的解释,或者至少解释得并不全面。生命的起源就是这样的一个例子。我们知道大约 40 亿年前,在原始汤之中出现了能够进行自我复制的分子,生命进化就从那里开始。但是,我们却不知道这些自我复制的分子最初是如何产生的。另一个例子是孤僻儿童往往具有非常好的记忆力这一现象。对于孤僻儿童的许多研究已经证实了这一点,但是至今还没有任何人成功地作出解释。

许多人相信,最终科学将能够解释这类事实。这似乎是一个相当合理的观点。分子生物学家们正致力于研究生命起源,只有悲观主义者才会说他们永远不会解决这一问题。诚然,问题并不简单，特别是，我们想要了解的是 40 亿年前地球上的环境条件。但即便如此，我们也没有理由认为生命的起源将永远无法解释。孤僻儿童的超强记忆力的例子也是这样。记忆力科学的研究仍处于初期，关于孤僻症的神经学基础仍有大量问题有待发现。显然我们不能保证这些问题最终一定可以得到解释。但是鉴于现代科学已经提出了大量成功的解释，认为今天许多有待解释的事实最终也能得到解释，必定是一个明智的判断。

但是这意味着原则上科学能够解释一切吗？或者说存在着某些一定永远能避开科学解释的现象吗？这不是一个容易回答的问题。一方面，断言科学能够解释一切似乎过于自负。另一方面，断言某个特定现象永远不能被科学地解释又似乎过于目光短浅。科学的发展和变化太迅速，从今天科学的观点来看似乎完全无法解释的现象，也许在明天就很容易解释了。

按照某些哲学家的观点，科学之所以永远不能解释一切，这有着纯逻辑学上的原因。为了解释某件事，无论它是什么，我们需要援用其他的事情。但是为第二件事提供解释的是什么呢？为了说明的需要，我们来回顾一下牛顿使用引力定律解释不同领域现象的例子。引力定律自身如何得到解释呢？如果有人问**为什么**所有的物体彼此都会产生引力作用，我们将如何作答？牛顿没有回答这一问题。在牛顿的科学中引力定律是一个基本原则：它解释其他事物，但自身无法得到解释。这一点给出的启示具有普遍意义。无论未来的科学能够解释多少事情，它给予的解释将不得不利用特定的基本定律和原则。任何事情都不能解释其本身，因此至少这些定律和原则中的一部分其自身无法获得解释。

无论怎样看待这种论证，无疑它都非常抽象。它意味着指出有些事情永远不能被解释，但并没有告诉我们它们是什么。然而，有些哲学家对于在他们看来科学永远不能解释的现象提出了具体的看法。其中一个例子就是意识——人类和其他高等动物这类会思考、能感知的生物所具有的典型特征。许多对意识之本质的研究已经并正在陆续由脑科学家、心理学家以及其他学者推进。但是近来的许多哲学家认为，无论这种研究探索到了什么，它将永远无法全面解释意识的本质。他们坚称，意识现象存在着固有的神秘，它们是再多的科学探索也不能去除的。

这种观点的根据是什么呢？基本的理由是：意识经验与世界上的其他任何事物都有根本的不同，它们有一个"主观层面"。例如，思考一下观看恐怖电影的经验。这是一种带有特殊"感受"的经验；在现代的术语中，拥有这种经验"似是而非"。也许有一天，神经科学家能够对使我们产生恐惧之感的复杂大脑活动给出一个详细的解释。但是，这将会解释为什么看恐怖电影就会产生这种感受，而不是其他某种感受吗？许多人认为不会。按照这种观点，对大脑的科学研究最多可以告诉我们哪些脑部程序是与哪些意识经验相联系的。这的确是使人感兴趣并有价值的信息。但是，它并没有告诉我们**为什么**带有特殊主观"感受"的经验由大脑的纯生理行为引发。因此，意识，或者至少它的一个重要方面，在科学上是无法解释清楚的。

尽管相当引人注目，这种论点还是非常有争议，并且不是所有的哲学家都认可，更不用说所有的神经科学家了。的确，1991 年出版的哲学家丹尼尔·丹尼特的一本闻名于世的著作就富有挑战性地冠名为《意识的解释》。支持意识在科学上无法解释这一观点的人有时就被斥为缺乏想象。即使当今的大脑科学不能解释意识经验的主观层面，难道我们就不能想象出现另一种完全不同的大脑科学，它拥有完全不同的解释工具，**的确**可以解释为什么我们的经验感受如此表现？一种源于哲学家的悠久传统试图告诉科学家，什么是可能的以及什么是不可能的，而后来科学的发展经常证明哲学家是错。是否同样的命运也在等待着那些认为意识必定无法接受科学解释的人，我们将拭目以待。

解释和还原

不同的科学学科是为了解释不同种类的现象被划分出来的。解释橡胶为什么不导电是物理学的任务，解释乌龟为何有如此之长的寿命是生物学的任务，解释较高的利率为什么可以削弱通货膨胀是经济学的任务，等等。总之，在不同学科之间有一个分工：每一科都致力于解释本领域的特定现象。这就解释了为什么学科之间通常不是互相竞争的关系——例如，为什么生物学家并不担心物理学家和经济学家会侵占他们的地盘。

尽管如此，人们却普遍认为科学的不同分支在地位上并不同等：有些分支要比其他分支更为根本。物理学通常被看做所有科学中最为根本的。为什么呢？因为其他学科所研究的对象最终都是由物理微粒构成的。以生物体为例。生物体是由细胞构成的，细胞本身是由水、核酸（如 DNA）、蛋白质、糖、脂类（脂肪）组成的，所有这些都是由分子或长分子链结合在一起构成的。而分子是由原子构成的，原子是物理学上的粒子。所以，生物学研究的对象最终就是非常复杂的物理学实体。同样的情况也适用于其他科学，甚至社会科学。以经济学为例。经济学研究的是市场上企业和消费者的行为,以及这种行为的后果。然而消费者都是人并且企业也是由人组成的;人是生物体,因此也是物理学实体。

这是否意味着物理学原则上能够包含所有更高层次的科学? 既然一切都是由物理学微粒构成的,如果我们有一门完整的物理学,它可以让我们精确地预测宇宙中每一个物理微粒的行为,那么其他所有的科学就一定会变得多余吗? 大多数哲学家都反对这种思路。毕竟,认为物理学有朝一日也许能够解释生物学和经济学所解释的事情,这似乎过于不切实际。直接从物理学规律推导出生物学和经济学规律的前景似乎太黯淡。无论未来的物理学有何进展,似乎都不可能预测经济的低迷。诸如生物学和经济学这样的科学非但不能被还原为物理学,而且似乎在很大程度上独立于它。

这导致了一个哲学上的难题。一门科学所研究的实体最终是属于物理学的,怎么会**无法**还原为物理学呢? 即使承认高阶的科学事实上独立于物理学,这种独立又是如何可能的呢? 按照一些哲学家的观点,问题的答案在于高阶科学研究的对象在物理学层面上是"被多重实现的"。为了解释多重实现的思想,我们不妨设想一个烟灰缸的集合。每一个个别的烟灰缸显然都是一个物理学实体,像宇宙中其他的每一个物体一样。但是烟灰缸的物理学组成却大不相同——有些也许是由玻璃做的,另一些也许是用铝做的,还有一些可能是塑料做的,等等。它们的尺寸、形状和重量可能也是不同的。烟灰缸可能具有的物理学属性在范围上实际上没有限制。因此不可能用纯物理学方式来定义"烟灰缸"这一概念。我们不可能找到一个正确的表达方式即"x 是一个烟灰缸当且仅当 x 是……"(空处由一个取自物理学语言的表达来填充）。这就意味着烟灰缸在物理学层面是被多重实现的。

　　哲学家经常援引多重实现来解释为什么心理学不能被还原为物理学或化学，而在原则上这一解释适用于任何高阶的科学。例如，我们来考察一下神经细胞比皮肤细胞寿命更长的生物学事实。细胞是物理实体，所以有人可能认为这一事实有一天会被物理学所解释。但是，细胞在微观物理学层面几乎肯定是被多重实现的。细胞虽然最终由原子构成，但是原子的精确排列在不同细胞中却会大不相同。所以，细胞的概念不可能用源自基础物理学的表达来定义。不存在这样的正确表达方式，即"x 是细胞当且仅当 x 是……"（空处由一个微观物理学语言的表达来填充）。如果这是正确的，就将意味着基础物理学永远不能解释为什么神经细胞比皮肤细胞寿命更长的问题，或者说事实上不能解释其他任何关于细胞的事实。细胞生物学词汇与物理学词汇并不能够以我们要求的方式一一对应。我们因此拥有了一个关于细胞生物学为什么不能还原为物理学的解释，尽管细胞是物理实体。并不是所有的哲学家都喜欢多重实现理论，但是该理论的确能够提供一种独立于高阶科学的巧妙的解释，从物理学方面和相互关系的方面来说都是如此。

第四章
实在论与反实在论

在哲学上被称为**实在论**和**观念论**的两种对立的思想流派之间一直存在着亘古的争论。实在论认为物理世界是独立于人的思维和感知而存在的。观念论否认这一点，认为物理世界以某种方式依赖于人的意识活动。对大多数人来说，实在论看似比观念论更为合理，因为实在论很好地契合了人们的常识看法，这种看法认为关于世界的事实是"在那里"等待着我们去发现的，而观念论却不以为然。的确，乍一看观念论好像相当可笑：既然人类消亡之后石头和树木大概仍将继续存在，在什么意义上能说它们的存在依赖于人的心智呢？事实上，实在论和观念论的争论要比这复杂得多，今天的哲学家们仍在继续讨论着。

传统的实在论／观念论论题属于一种被称为形而上学的哲学领域，但实际上这一论题和科学之间并没有任何特别的关系。本章我们所关注的是一种更为现代的争论，它与科学特别相关并且在某些方面类似于传统的论题。这一争论发生在一种被称为**科学实在论**的立场及其对立面**反实在论**或**工具论**的立场之间。从现在开始，我们将使用"实在论"一词指称科学实在论，用"实在论者"指称科学实在论者。

科学实在论与反实在论

像大多数哲学"主义"一样,科学实在论①是以不同的版本出现的,我们很难用一种完全精确的方法对其加以定义。但其基本观点却是显明的。实在论者认为科学的目的就在于为世界提供一种正确描述。这听起来像是一种无关紧要的学说,因为当然没有人会认为科学意在提出一种关于世界的错误描述。但是,这却不是反实在论者所思考的。相反,反实在论者认为科学的目的在于对世界的某个特定的**部分**——"可观察"的部分——提供一种正确描述。至于世界"不可观察"的部分,按照反实在论者的观点,科学所描述的是真是假都无关紧要。

对于世界的可观察部分,反实在论者确切指称的是什么呢?他们指称桌子、椅子、树木和动物、试管和本生灯、雷雨和下雪等等的日常世界,诸如此类的事物都是可以被人们直接感知的——这也就是我们称之为可观察的意思。科学的一些分支就是专门处理可观察对象的,古生物学,或者说对化石的研究,就是一个例子。化石是容易观察到的——任何具有正常视力的人都能够看到它们。但是,其他的一些科学分支却是探究不可观察的实在领域的,物理学就是一个明显的例子。物理学家们不断提出关于原子、电子、夸克、轻子以及其他奇异粒子的理论,所有这些粒子在常规世界中都不能被观察到。这种类型的实体已经超出了人类观察能力的可及范围。

关于古生物学之类的科学,实在论和反实在论之间并没有意见分歧。就化石的研究而言,因为化石是可观察的,实在论者关于科学意在真实地描述世界的论题和反实在论者关于科学

① "实在论"原文 realism ,"主义"原文 ism。——编注

意在真实地描述可观察世界的论题显然是一致的。但是涉及到物理学一类的科学时，实在论和反实在论就出现了分歧。实在论者认为，当物理学家提出关于电子和夸克的理论时，他们是在试图提供对于亚原子世界的真实描述，就像古生物学家试图提供对于化石世界的真实描述一样。反实在论者不同意这种观点，他们在亚原子物理学和古生物学理论之间看到了一种根本的区别。

当反实在论者们谈论不可观察的实体时，他们会认为物理学家**正在**研究什么呢？通常，他们会认为这些实体仅仅是顺手虚构的，是物理学家们为了预测可观察的现象而提出的。例如，来看一下气体动力学理论，这一理论认为任何体积的气体都包含了大量处于运动中的微小实体。这些实体——分子——是不可观察的。通过动力学理论，我们能够推知关于气体可观察行为的各种结论。例如，如果气压保持不变，加热标本气体就会导致气体膨胀，这一点可以通过实验证实。按照反实在论者的观点，在气体动力学理论中假定不可观察实体的唯一目的就是推出这类结论。气体是否**的确**包含了运动的分子，这一点并不重要；动力学理论的要义不在于真实描述隐藏的事实，而在于提供一种预测观察数据的方便途径。我们可以看出为什么反实在论有时被称为"工具论"——它认为科学理论是有助于我们预测观察数据的工具，而不是描述实在之潜在本质的努力。

由于实在论／反实在论的争论涉及到科学的目标，人们可能会认为这一争论仅仅靠询问科学家本人就可以得到解决。为什么不在科学家当中做一个民意测验问问他们自己的目标呢？然而这一建议并未切中要义——它过于从字面来理解"科学的目标"一词。当我们问什么是科学的目标时，并不是要问各个科学家的目标。确切地说,我们是要问如何最好地弄清科学家所说的和所做的——如何解释科学事业。实在论者认为我们应该将所有的科学理论解释为对实在的尝试性描述;反实在论者认为这一解释对于谈论不可观察的实体和过程的理论来说是不恰当的。揭示出科学家们自己对实在论 / 反实在论之争的看法肯定是一件有趣的事,然而,这一问题终究还是一个哲学问题。

反实在论的许多动机都来源于如下信念:我们实际上不能获得关于实在的不可观察部分的知识——这种知识超出了人类的知识视域。按照这一观点,科学知识的界限是由我们的观察能力设定的。所以,科学能够给予我们关于化石、树木、冰糖的知识,但是不能给予我们关于原子、电子和夸克的知识——后者都是不可观察的。这一观点并非完全不合情理。没有人会真正怀疑化石和树木的存在,但是却可能去怀疑原子和电子的存在。正如我们在本书最后一章将要看到的,19 世纪后期的许多重要科学家都怀疑原子的存在。如果科学知识限于可被观察的范围内,那么所有接受这一观点的人,显然都必须解释**为什么**科学家会提出关于不可观察的实体的理论。反实在论者给出的解释是,这些理论都是顺手虚构的,是为了预测事物在可观察世界中的表现而提出的。

实在论者不认为科学知识受限于我们的观察能力。相反,他们认为我们已经切实了解了不可观察的实在,因为存在着种种理由认为我们最好的科学理论都是正确的,而最好的科学理论都在探讨不可观察的实体。例如,来看一下物质的原子论,这一理论声称所有的物质都是由原子构成的。原子论能够解释有关世界的大量事实。按照实在论者的说法,这是表明这一理论正确的很好的证据,也就是说,物质的确是由如这一理论所描述的以某种方式表现的原子构成的。尽管存在着支持该理论的明显证据,该理论当然**可能**仍是错误的,但所有的理论都可能是这样。仅仅因为原子不可观察,还没有理由不将原子论解释成一种对实在的尝试性描述——多半是一种成功的描述。

严格来说我们应该区分两种反实在论。根据第一种反实在论,根本不能从字面上去理解对不可观察实体的讨论。因此,(例如)当一位科学家提出一种关于电子的理论时,我们不应该认为他是在宣称被称为"电子"的那种实体的存在。相反,他谈论电子是比喻性的。这种反实在论在 20 世纪上半叶非常流行,但是现在已经很少有人提倡了。它主要是受到语言哲学中的一种学说的推动。按照这种学说,对原则上不可观察的事物作出有意义的断言是不可能的——当代很少有哲学家还接受这一学说。第二种形式的反实在论承认,关于不可观察的实体的探讨应该从字面来理解:如果一个理论认为电子带负电荷,当电子存在并且带负电荷时这一理论为真,反之为假。但是反实在论者声称,我们将不会知道到底是哪一种情形。因此,对待科学家关于不可观察的实在所提出的主张,正确的态度应该是一种彻底的不可知论。这些主张要么正确要么错误,但是我们不能辨别其正误。现代的大多数反实在论都属于第二种形式。

"无奇迹"说

许多假设不可观察实体存在的理论**在经验上是成功的**——它们对可观察世界中物体的表现作出了准确的预测。上面提到的气体动力学理论就是一个例子,还有许多这样的例子。此外,这类理论通常还具有重要的技术用途。例如,激光技术所依据理论的基础,就与一个原子中的电子从高能状态转变为低能状态过程中发生的情况相关。激光技术不断发挥作用——使我们能够矫正视力、用导弹袭击敌人，以及做其他很多事。因此奠定激光技术之基础的理论在经验上是高度成功的。

假设不可观察实体存在的那些理论在经验上的成功，是支持科学实在论的一种最有力的论点，被称为"无奇迹"说。按照这种说法，如果一个探讨电子和原子的理论精确预测了可观察的世界，那么，除非实际上电子和原子真的存在，否则这一理论就将是一种罕见的巧合。如果不存在原子和电子，怎么解释理论与观察数据的紧密相符？类似地，除非假设相关理论是正确的，我们又该怎样去解释理论推动了技术的进步？如果正如反实在论者所言，原子和电子仅仅是"顺手虚构的"，为什么激光能有效地发挥作用？按照这一观点，成为一个反实在论者就等于相信奇迹。既然在存在非奇迹的替代情形时显然最好不要相信奇迹，我们就应该成为实在论者而不是反实在论者。

"无奇迹"说并不是想**证明**实在论是正确的而反实在论是错误的。确切地说，它是一个合理性论点——一种最佳解释推论。待解释的现象是这样一种事实，即许多假设不可观察实体存在的理论在经验上高度成功。"无奇迹"说的倡导者们声称，对于这一事实的最佳解释就是这些理论是正确的——相关的实体的确存在，正如理论所说的那样以某种方式表现出来。除非接受这种解释，否则我们的理论在经验上的成功就仍是一个无法解释的谜团。

反实在论者用不同的方式回应"无奇迹"说。其中一种回应诉诸关于科学史的某些事实。在历史上有许多理论，我们现在认为那些理论是错误的而在当时它们在经验上却相当成功。在一篇著名的文章中，美国科学哲学家拉里·劳丹从不同科学门类和科学时期中抽取、列举了不下 30 种这类理论。燃烧的燃素说就是一个例子。这一理论直到 18 世纪末还广泛流行,它主张任何物体燃烧时都会释放一种叫做"燃素"的物质进入空气中。现代化学告诉我们这一理论是错误的:不存在类似燃素的物质。相反,燃烧发生于物体在空气中与氧气发生化学反应时。尽管燃素是不存在的,燃素说在经验上却相当成功:它非常恰当地符合当时能获取的观察数据。

　　这类例子表明,支持科学实在论的"无奇迹"说有些操之过急。"无奇迹"说的支持者们将当今科学理论的经验性成功看做是证明了这些理论的正确性。但是科学史表明,在经验上成功的理论通常被证明是错误的。所以我们如何知道同样的命运不会降临今天的理论? 例如,我们如何知道物质的原子论不会和燃素说一样?反实在论者声称,一旦对科学史投入应有的关注,我们就能看到从经验性成功到理论正确性的推导是不可靠的。因此,对待原子论的理性态度应是一种不可知论——它可能正确,也可能不正确。反实在论者说,我们反正无从得知。

　　这是对"无奇迹"说的一个强有力的反击,但不是决定性的一击。一些实在论者通过稍微修正这一说法来加以回应。按照改进后的说法,一种理论的经验性成功证明的是:理论就不可观察的世界所述的内容是近似为真,而不是准确地为真。这一弱化的陈述更容易抵御来自科学史上反例的攻击。同时它也更为温和:它允许实在论者承认今天的理论在某个细枝末节上可能不正确,但仍坚称今天的理论大致是正确的。另一种修正"无奇迹"说的方法是改进经验成功这一概念。一些实在论者认为,经验成功不仅仅是与已知的观察数据相符的问题,它还能使我们去预测新的尚属未知的观察数据。与经验成功的这一更严格的标准相比较，找出在经验上成功、后来却被证实为错误的史例就更不容易了。

尚不确定这些改进能否真的挽救"无奇迹"说。它们当然减少了历史上的反例数量，却没有完全消除。仍旧存在的一个反例是1690年首次由克里斯蒂安·惠更斯提出的光的波动理论。按照这一理论，光是由"以太"这种不可见的介质中的波状振动构成的，而以太被认为充满着整个宇宙。（波动说的竞争理论是牛顿支持的光的微粒说，微粒说坚持光是由光源放出的极小微粒构成的。）直到法国物理学家奥古斯丁·菲涅耳在1815年用公式表示了光的波动理论的数学形式，并将其用以预测一些令人惊奇的新光学现象，这一理论才被广为接受。光学实验证实了菲涅耳的预测，并且使许多19世纪的科学家相信光的波动理论是正确的。然而，现代物理学又告诉我们这一理论并不正确：并不存在类似以太这样的物质，所以光并不是由以太中的振动组成。我们再次碰到一个错误的但在经验上成功的理论。

这一例子的重要特征是，即使是改进后的"无奇迹"说，同样可以被推翻。菲涅耳的理论**的确**作出了新颖的预测，所以即使就经验成功的更为严格的标准来说，它也有资格被认为是经验上成功的。而且很难理解，既然菲涅耳的理论建基于并不存在的以太概念之上，它如何能被称做"近似为真"。无论声称一个理论近似为真的确切所指是什么，一个必然条件一定是这一理论所谈论的实体的确存在着。简言之，即使按照一种严格的对经验成功概念的理解，菲涅耳的理论在经验上也是成功的，但却不是近似为真的。反实在论者说，这一例子的寓意是我们不应该仅仅因为现代科学理论在经验上如此成功就假设它们大致正确。

因而,"无奇迹"说对于科学实在论来说是否是一个好的论点,这一点尚不确定。一方面,正如我们看到的,这一论点面临着相当严肃的反对意见。另一方面,关于这一论点也存在着一些在直觉上引人注目的东西。当人们考虑到那些假设了原子、电子等实体存在的理论令人惊异的成功时,就很难接受原子和电子不存在。但是正如科学史所表明的,无论当今的科学理论如何与观察数据相符,我们都应该对认为这类理论正确的假设持谨慎态度。过去许多人都曾作过上述假设,却被证明是错误的。

可观察／不可观察的区分

实在论和反实在论之争的核心是可观察事物和不可观察事物之间的区分。到目前为止,我们只是认为这一区分是当然的——桌子和椅子是可观察的,而原子和电子是不可观察的。然而,这种区分在哲学上其实是有问题的。事实上,科学实在论的一种主要观点认为,以一种原则性的方式在可观察／不可观察之间作出区分是不可能的。

为什么这一观点出自科学实在论?因为反实在论的一致性主要依赖于可观察和不可观察之间的明显区别。回想一下:反实在论者倡导对科学主张持不同的态度,视这些科学主张是关于实在的可观察部分还是不可观察部分而定——我们仍应对后者而非前者的正确性持不可知论的态度。反实在论由此预设我们能够将科学主张分为两类:关于可观察实体、过程的科学和关于不可观察实体、过程的科学。如果事实是不能用必要的方式作出这种分类,反实在论显然会陷入严重的困境之中,实在论不战而胜。为什么科学实在论者通常热衷于强调与可观察／不可观察区分相关的问题？原因就在这里。

这类问题中有一个涉及到观察和检测之间的关系。类似于电子这样的实体显然在常规意义上是不可观察的,但是它们的存在可以通过被称为粒子检测器的特殊仪器来检测到。最简单的粒子检测器是云室,它由一个充满着空气和饱和蒸汽的密闭容器构成(参见图9)。当带电粒子(如电子)穿过云室时,它们就会与空气中的中子相碰撞,将中子转化为离子;水蒸汽在这些可以导致液滴形成的离子周围凝结,这一切可以通过肉眼看到。我们可以通过观察这些液滴的轨迹来追踪电子在云室中的路径。这是否意味着电子终究能被观察到? 大多数哲学家会说不:云室能使我们检测到电子,但不是直接观察到它们。这就如同,高速喷气式飞机可以通过其蒸汽留下的轨迹被检测到,但观察到轨迹并不等于观察到飞机本身。然而,观察和检测之间的区分通常很明显吗? 如果不是,反实在论者的立场就陷入困境之中。

在20世纪60年代对科学实在论的一个著名辩护中,美国哲学家格罗弗·马克斯韦尔针对反实在论者提出以下问题。考虑一下下述事件的顺序:用肉眼看某些东西,透过窗子看某些东西,借助一副高度数眼镜看某些东西,借助双筒望远镜看某些东西,借助一个低倍显微镜看某些东西,借助一个高倍显微镜看某些东西,等等。马克斯韦尔认为,这些事件取决于一个平稳的连续体。那么,我们如何来决定哪些行为是观察,哪些不是?生物学家能够借助高倍显微镜来观察微生物吗?或者说,他只能用与物理学家在云室中检测电子存在一样的方法来检测微生物的存在吗?如果某些东西仅仅在借助精密科学仪器的情况下才能被看到,它们应被视为可观察的还是不可观察的? 在

图 9 第一组照片中的一张显示了亚原子粒子在云室中的轨迹。1911
年,英国物理学家、云室的发明者 C.T.R.威尔森在剑桥的卡文迪什实
验室拍下这张照片。嵌入在云室中的金属舌片顶部的少量镭放射的 α
粒子产生了这一轨迹。带电荷的粒子在云室中沿着蒸汽移动,使气体
电离;水滴凝结在离子上,从而产生粒子经过了的液滴的轨迹。

我们拥有检测的例子而不是观察的例子之前,仪器制造能达到何种精密程度呢?马克斯韦尔认为,没有一种原则性的方法来回答这些问题,所以反实在论者将实体分为可观察的和不可观察的这种尝试注定要失败。

马克斯韦尔的论证得到如下事实的支持:科学家们自己有时借助于精密的仪器来谈论"观察"粒子。在哲学文献中,电子通常被认为是不可观察实体的范例,但是科学家们却常常津津乐道于通过粒子检测器来"观察"电子。当然,这一点并不能证明哲学家们错了以及电子终究是可观察的,因为科学家的谈论可能最好被理解为"说说罢了(facon-de-parler)"。类似地,正如我们在第二章中看到的,科学家谈论一种理论具有"实验证据"也不意味着实验就真的能证明该理论是正确的。然而,如果正如反实在论者所言,真的存在着一种哲学上重要的可观察/不可观察的区分,很奇怪它竟与科学家自身说话的方式如此背离。

马克斯韦尔的论证是有力的,但决不是完全决定性的。当代一位重要的反实在论者范·弗拉森认为,马克斯韦尔的观点仅仅表明"可观察的"是一个模糊概念。模糊概念是指它有处于边界线上的情形——不能清晰地归入或不归入其中的情形。"秃子"就是一个明显的例子。因为掉头发是渐渐发生的,很难说许多人到底是不是秃子。但是范·弗拉森却指出,模糊概念完全可用,并且能够表明世界上的真正差别。(实际上,大多数概念至少在某种程度上都是模糊的。)没有人会仅仅因为"秃子"这个词是模糊的,就坚持认为秃子和有头发的人之间的区别是非实在的或不重要的。可以肯定的是,如果我们试图在秃子和有头发的人之间划出一个明显的界线,这种界线会有任意的成分。但是,因为存在着秃子和不是秃子的清晰的例子,无法划出这种明显的界线就是无关紧要的。尽管概念存在着模糊性,但它却可以很好地使用。

按照范·弗拉森的观点,同样的情况也完全适用于"可观察的"。明显存在着能观察到的实体的例子,例如椅子;也明显存在不能观察到的实体的例子,例如电子。马克斯韦尔的观点强调存在着处于边界线上的情形的事实,其中我们不能确定相关的实体是否能被观察到或仅被检测到。所以,如果我们试图在可观察和不可观察的实体间划出明确的界线,这一界线就不可避免地会有些武断。但是正如秃子的例子一样,无论如何它并不表明可观察/不可观察的区分是不真实或不重要的,因为在可观察/不可观察两边都存在着清晰的例子。所以范·弗拉森认为,"可观察"一词的模糊性对于反实在论者来说并不是什么麻烦。它仅仅是对准确性设定了一个上限,反实在论者能借助这一上限来陈述立场。

这一观点有多少说服力?范·弗拉森认为边界情形的存在以及随之产生的无法客观地划出明显界限的结果,并不能表明可观察/不可观察的区分是非实在的,这一观点当然是正确的。就这一点来说,范·弗拉森的观点成功反驳了马克斯韦尔。然而,表明在可观察和不可观察的实体间存在着真正的区分是一回事,表明这种区分能够担当反实在论者希望负载其上的哲学任务又是另一回事。回想一下反实在论者倡导对关于实在之不可观察部分的论述持完全不可知论的态度——他们说,我们无法知道这些论述是真还是假。即使我们承认范·弗拉森的观点是对的,即存在着关于不可观察实体的明显例子,并且这一观点也足以使反实在论者继续论证其立场,反实在论者仍然需要为如下想法提供理由：关于不可观察实在的认知是不可能的。

不充分论证说

有一种支持反实在论的观点主要关注科学家的观察数据与他们的理论主张之间的关系。反实在论者强调，科学理论所要符合的最终数据在特性上总是可观察的。(许多实在论者都会同意这一论断。)为了表明这一点，我们再来思考一下气体的动力学理论，它声称任何气体都是由处于运动中的分子构成的。因为这些分子都是不可观察的，显然我们不能通过直接观察各种气体标本来检验这一理论。相反，我们需要从理论中推导出一些能被检验的陈述，这些陈述总是关于可观察实体的。正如我们所见，动力学理论暗含了如果气压保持不变，气体受热就会膨胀。通过在实验室里观察相关仪器的读数我们可以直

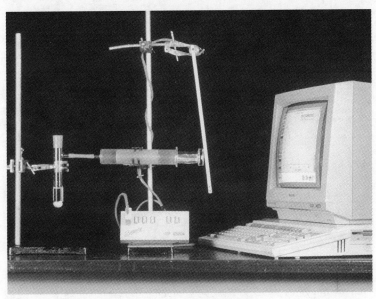

图 10 测量气体体积随着温度改变而变化的膨胀仪。

接检验这一陈述(见图10)。这一例子解释了一个普遍的事实：观察数据构成了关于不可观察实体的论断的最终证据。

反实在论者于是认为观察数据"未充分论证"科学家们在此数据基础上提出的理论。这是什么意思呢？它意味着观察数据原则上能够由许多不同的、相互不兼容的理论加以解释。在动力学理论的例子中反实在论将会声称，这种观察数据的**一种**可能的解释是，正如动力学理论所述的那样，气体包含着大量处于运动中的分子。但是他们也认为，还存在着其他可能的解释，这些解释与动力学理论相冲突。所以按照反实在论者的观点，假设存在着不可观察实体的科学理论是由观察数据不充分论证的——总是存在着大量同样能够很好地解释观察数据的竞争理论。

很容易就能理解，为什么不充分论证说支持反实在论的科学观。因为，如果理论总是由观察数据不充分论证的，我们如何能有理由相信某个特定的理论是正确的？假设一位科学家支持某种关于不可观察实体的既有理论，理由是这一理论能够解释大量观察数据。反实在论的科学哲学家走上前来，宣称观察数据实际上能被各种替代理论所解释。如果反实在论者是正确的，那么就将得出科学家对于其理论的信心放错了地方。因为什么原因科学家要选择她所选的理论，而不是选择另一种理论呢？在这样一种情形中，科学家真的应该承认她自己也不知道哪种理论正确吗？不充分论证自然导致反实在论的结论，即不可知论是面对关于不可观察的实在领域的主张时所应持的正确态度。

但是，是否真如反实在论者所断言的，一组特定的观察数据总是能被许多不同的理论加以解释？实在论者通常认为这一论断仅仅在琐碎和无趣的意义上才是正确的,借此来回应不充分论证说。从原则上说,对于某组特定的观察数据总是存在着不止一种可能的解释。但是实在论者认为,这并不能得出所有这些可能的解释都一样好。两个理论都能解释我们的观察数据并不意谓着在它们之间就无法选择。比如,一种理论可能就比另一种更简单,或者它用一种在直觉上更合理的方式来解释数据,或者它可能假设了更少的隐性原因,等等。一旦我们承认,除了与观察数据的兼容性外还存在着别的理论选择的标准,不充分论证问题就会消失。并不是所有对观察数据的可能解释彼此都是一样好的。即使动力学理论所解释的数据原则上能由其他替代理论来解释,也不能得出这些替代理论就能和动力学理论解释得一样好。

这种对不充分论证说的回应得到了以下事实的支持:科学史上很少有不充分论证的真实情形。如果正如反实在论者所言,观察数据总是能被许多不同的理论同样好地加以解释,我们就真的会看到科学家们处在永远地相互争论之中吗?这并不是我们见到的实际情形。事实上,当我们察看历史记载时,情况几乎和不充分论证说使我们期望的正好相反。科学家们远非面对大量对于其观察数据的可能解释,他们通常甚至难以找到一**种**与数据充分符合的理论。这就支持了实在论者的观点,即不充分论证说仅仅是一种哲学家的担忧,它与实际的科学实践没有多大关系。

反实在论者不可能受这一回应的影响。毕竟,哲学的担忧仍然是真实的担忧,即使它们的实践意义微乎其微。哲学可能改变不了世界,但是这并不意味着它不重要。而类似于简单性这样的标准能被直接用于对两个竞争理论的判定,这种看法直接引起了如下的棘手问题:为什么更简单的理论应该被认为更可能是正确的;我们在第二章中涉及了这一问题。反实在论者通常承认,在实践中,通过使用类似简单性的标准去辨别对观察数据的两个竞争性解释,不充分论证问题就能被消解。但是他们否认这类标准是正确性的可靠标志。更简单的理论操作起来可能更简便,但是它们不是本质上就比复杂的理论更站得住脚。所以不充分论证说坚持认为:对于观察数据总是存在着多种解释,我们无法知道哪种理论是正确的,所以关于不可观察实在的知识是不可能获得的。

　　然而,争论并没有到此结束;还有一种来自实在论者的更进一步的反驳。实在论者指责反实在论者选择性地运用不充分论证说。实在论者声称,如果这一说法是自始至终被运用的,它就不仅会取消不可观察世界的知识,还会取消大部分可观察世界的知识。为了理解实在论者为什么这么说,需要注意:许多可观察的事物实际上从来没被观察到过。例如,行星上绝大多数的生命体从来都不曾被人类观察到,这些生命体显然是可观察的。或者想想类似于巨大的陨星撞击地球这样的事件。没有人曾目击过这类事件,但是它们显然也是可观察的。它正好在没有人生存的那个地点和那个时间发生。在可观察的事物中,只有一小部分事实上被观察到。

　　关键之处正在于此。反实在论者宣称,实在的不可观察部分超出了科学知识的界限。所以他们承认,我们能够拥有关于可观察的但**尚未**观察到的物体和事件的知识。但是,关于尚未观察到的物体和事件的理论与关于不可观察物体和事件的理论一样,都是被我们的观察数据不充分论证的。例如,假设一位科学家提出如下假说:一陨星在 1987 年撞击月球。他列举了各种观察数据来支持该假说,比如,月球的卫星照片显示了一个在 1987 年前不存在的大坑。然而,这一观察数据原则上能由许多替代假说来解释——可能是一次火山爆发造成了大坑,或是一次地震。也可能拍下卫星照片的照相机出了毛病,根本就不存在大坑。所以,科学家的假说是由观察数据不充分论证的,即使这一假说是关于一个完全可观察的事件——陨星撞击月球。实在论者声称,如果一以贯之地运用不充分论证说,我们就会被迫得出如下结论:我们仅仅能获得关于实际上已经被观察到了的事物的知识。

这一结论极不合理,并非任何科学哲学家都乐意接受。科学家告诉我们的大量事实都涉及尚未被观察到的事物——想想冰川时期、恐龙、大陆漂移,等等。说关于尚未观察到之物的知识是不可能的,就等于说大多数科学知识完全不是知识。当然,科学实在论者并不接受这一结论。相反,他们把它视为证明不充分论证说错误的证据。因为,尽管存在着关于未观察到之物的理论由观察数据不充分论证这一事实,很明显科学仍给了我们关于尚未观察到之物的知识,这就推导出不充分论证并不是获取知识的障碍。因此,关于不可观察之物的理论也是由观察数据不充分论证的这一事实,并不意味着科学不能给我们提供关于世界的不可观察领域的知识。

事实上,以这一方式提出主张的实在论者是在声称,不充分论证说提出的问题仅仅是归纳问题的一个复杂版本。说一种理论是由观察数据不充分论证的,等于说存在着能够解释同样数据的替代理论。但是这实际上等于说数据并不必然推出理论:从观察数据到理论的推导是非演绎的。无论理论是关于不可观察实体的还是关于可观察但尚未观察到的实体的,这都没有区别——这两类情形中的逻辑是一样的。当然,表明不充分论证说仅仅是归纳问题的一种版本并不意味着它能被忽视。正如我们在第二章中看到的,关于归纳问题应如何处理这一点并没有形成共识。然而这也不意味着对于不可观察的实体就不存在**特殊的**困难。因此,实在论者声称,反实在论者的立场终究是武断的。在理解科学如何给我们提供关于原子和电子的知识时,无论存在着什么样的问题,都与在理解科学如何为我们提供关于常规、普通对象的知识时遇到的问题一样。

第五章

科学变迁和科学革命

　　科学思想变化迅速。事实上挑出任意一门你喜欢的科学学科，你都能确信那门学科中的流行理论已和 50 年前的大不一样，和 100 年前的更是完全不同。与哲学和人文学科等其他的智识活动相比，科学是一个快速变迁的领域。大量有趣的哲学问题都聚焦于科学变迁。科学观念随着时间不断地变化，是否存在着一种清晰的变迁方式呢？当科学家们放弃现有理论而支持一种新的理论时，我们该作何解释？最新的科学理论在客观性上是否就比先前的更好？客观性的概念是否就有意义呢？

　　大多数关于这些问题的现代讨论都源于已故美国科学史学家和科学哲学家托马斯·库恩的一部著作。1963 年库恩出版了《科学革命的结构》一书，它无疑是过去 50 年中最有影响的科学哲学著作。库恩思想的影响已经渗透到社会学和人类学等其他学科中，甚至广泛渗透到一般的精神文化之中。(《卫报》将《科学革命的结构》列为 20 世纪最有影响的 100 本书之一。)为了理解库恩的思想为什么引起如此轰动，我们需要简要回顾库恩的书出版之前科学哲学的发展状况。

逻辑实证主义的科学哲学

战后英语世界占统治地位的哲学思潮是逻辑实证主义。最初的逻辑实证主义者是 20 世纪 20 年代以及 30 年代初，在莫里茨·石里克领导下，由一群在维也纳相遇的哲学家和科学家组成的松散团体。(我们在第三章提到的卡尔·亨普尔与实证主义者交往密切，卡尔·波普尔也一样。)为了逃避纳粹的迫害，大多数实证主义者移民去了美国，在那里他们和追随者们一直对学院派哲学产生着强大的影响，直到大约 60 年代中期之后，这一哲学思潮开始解体。

逻辑实证主义者对自然科学、数学和逻辑高度重视。在 20 世纪初的几年里人们见证了激动人心的科学进步，特别是物理学领域的，这极其深刻地影响了实证主义者。实证主义者的目标之一就是使哲学本身变得更为"科学"，以使哲学领域也出现类似的进步。就科学来说，对实证主义者影响特别深的是它表面上的客观性。实证主义者相信其他的领域更多地表现探究者的主观意见，而科学问题能够用一种完全客观的方式解决。实验检验一类的方法使科学家能够将理论直接和事实相比较，从而作出一个基于可靠信息的、无偏见的关于理论价值的评价。因此，对于实证主义者来说科学是一种范式性的理性活动，一条通向既存真理的最可靠的道路。

尽管实证主义者高度尊重科学，他们却很少关注科学史。事实上，实证主义者认为哲学家们从科学史的学习中获益很少。这主要是因为，他们在所谓"发现的语境"和"证明的语境"之间作出了严格的区分。发现的语境指的是科学家获得一个特定理论的实际历史过程。辩护的语境则指理论已经存在时，科

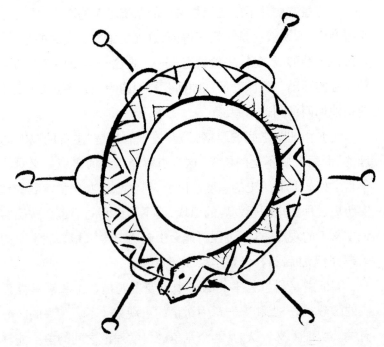

图 11 凯库勒在一场梦后偶然想到苯的六边形结构假说，梦里凯库勒看见一条蛇正试图咬住自己的尾巴。

学家证明他的理论所使用的方法——包括检验理论，寻找相关证据，等等。实证主义者认为前者是一种主观的、心理的过程，不受精确规则的支配，而后者是一个客观的逻辑问题。他们主张，科学哲学家应该致力于研究后者。

一个例子有助于使这一观念更为清晰。1865 年比利时科学家凯库勒发现苯分子具有六边形的结构。表面上看，他是在一场梦后偶然想到苯的六边形结构假说，在那场梦里凯库勒看见一条蛇正咬住自己的尾巴（见图 11）。当然，事后凯库勒还要科学地检验梦醒后提出的假说，他也这么做了。这是一个极端的例子，但它表明科学假说能够通过看似最不可能的方法获得——这些方法并不总是深思熟虑的系统思考的产物。实证主义者会说,假说最初是如何获得的,这无关紧要。要紧的是,一旦假说已经形成,怎样去检验它——正是这一点使科学成为一种理性的活动。凯库勒最初如何获得他的假说并不重要,重要的是他如何证明这一假说。

发现和证明之间的严格区分,关于前者是"主观的"和"心理的"而后者则不是的信条,这两点解释了为何实证主义对科学哲学持如此非历史的态度。因为,科学思想变化和发展的真实历史过程完全源于发现的语境,而不是证明的语境。按照实证主义者的观点,这一过程可能会引起历史学家和心理学家的兴趣,但无法带给科学哲学家任何东西。

实证主义科学哲学的另一个重要主题是理论和观察性事实的区分;这一点和上一章中讨论的可观察/不可观察的区分有关。实证主义者认为,两个竞争的科学理论之间的争论能够用一种完全客观的方法——将理论和"中立的"观察性事实直接比较,这种方法任何一方都能接受——来解决。实证主义者之间在如何准确描述这些中立的事实这一点上意见不一,但是他们都坚定地认为这些事实是存在着的。没有理论和观察性事实之间明晰的区分,科学的合理性和客观性将变得折中,而实证主义者坚定地认为科学是理性的和客观的。

科学革命的结构

库恩是一位训练有素的科学史家,他确信哲学家们能够从科学史的学习中获益良多。他认为,对科学史的重视不足使实证主义者得到的是一种关于科学事业的不准确和幼稚的图景。正如他书的标题所表明的,库恩尤其对科学革命——现存科学思想被新的思想彻底代替的剧变时期——感兴趣。科学革命的例子包括天文学中的哥白尼革命、物理学中的爱因斯坦革命以及生物学中的达尔文革命。每一次革命都导致了科学世界观的根本变化——一系列现存的思想被另一些完全不同的思想所推翻。

当然，科学革命还是相对较少地发生的——大多数时间任何特定的科学都不处于革命状态。库恩创设了"常规科学"这一术语，来描述当科学家所属的学科没有经历革命性的变化时他们所从事的每天平常的科学活动。库恩对常规科学进行解释的核心概念是**范式**。范式包括两个组成部分：首先，某一科学共同体的所有成员在某一特定时期都能接受的一系列基本的理论假设；第二，由上述理论假设解决了的、出现在相关学科教科书上的一系列"范例"或特定科学问题。但是，范式不仅仅是一个理论(尽管库恩有时交换使用这两个词)。当科学家们共用一个范式时，他们并不仅仅赞同特定的科学命题，他们还在自己所属领域的未来科学研究应该如何推进、哪些是相关的需要解决的问题、解决那些问题的恰当方法是什么、那些问题的可接受解决办法应该如何等问题上意见一致。简而言之，一个范式就是对科学的总体观点——联结科学共同体并且允许常规科学发生的一系列共享的假设、信念和价值观。

常规科学准确地讲包括什么呢？按照库恩的观点，常规科学主要是一种**解惑**的活动。无论一个范式多么成功，它都将遇到特定的困难——那些它无法涵盖的现象、理论预见和实验事实之间的龃龉，等等。常规科学家的工作就是试图消除这些较小的困惑，同时使得对范式的改变尽可能少。所以常规科学是一种相当保守的活动——它的研究人员不是试图作出任何惊天动地的发现,而仅仅是要发展和扩充既存的范式。用库恩的话说,"常规科学并不试图去发现新奇的事实或发明新理论,成功的常规科学研究并不会发现新东西"。最重要的是,库恩强调常规科学家并不试图**检验**范式。相反,他们不加疑问地接受范式,并在范式所设定的范围内开展研究。如果一位常规科学家得到了一个有悖于范式的实验结果,她通常会假定实验方法有误,而不认为是范式错了。范式本身是不可商榷的。

常规科学的时期一般能持续几十年,有时甚至是几个世纪。在此期间科学家们逐渐地阐释范式——恰当调整范式,充实细节,解答越来越多的困惑,扩大范式的适用范围,等等。但是随着时间的推移,出现了**反常**——那些不论常规科学家们如何努力都无法与范式的理论假设相一致的现象。当反常在数量上还很少的时候,它们容易被忽视。但当反常累积得越来越多时,一种逐渐增强的危机感就笼罩着科学共同体。对既存范式的信心瓦解了,常规科学的进程也暂时趋停。这标志着库恩所说的"革命的科学"时期的开始。在此时期,主要的科学观念都**处于**公开竞争的地位。各种对旧范式的替代方案被提出,最终,一种新的范式就被确立。大约需要一代人的时间,科学共同体的所有成员都转而信奉新范式——这标志着科学革命的完成。因此,科学革命的本质就是从旧的范式转向一种新的范式。

库恩将科学史概括为被偶尔的科学革命中断的漫长常规科学时期,这得到了许多哲学家和科学史家的响应。来自科学史上大量的例子恰好符合库恩的概括。例如,当我们考察从托勒密到哥白尼的天文学变化,或从牛顿到爱因斯坦的物理学变化时,库恩所描述的许多特征都在其中显现出来。坚持托勒密体系的天文学家们都共有一种范式,这一范式建立在地球静处于宇宙中心的理论之上，为这些天文学家们的研究搭建了一个不受质疑的背景。在 18、19 世纪坚持牛顿体系的物理学家们也是如此，他们的范式建立在牛顿的力学和引力理论之上。在这两个例子中，库恩关于旧范式怎样被新范式取代的解释相当准确地得以适用。也有一些科学革命并非如此精准地符合库恩模型，例如近来生物学上的分子革命。然而尽管如此，大多数人都赞同，库恩对于科学史的描述蕴含着重要的价值。

为什么库恩的思想能引起如此的风暴？因为除了对科学史纯粹的描述性陈述外，库恩还提出了一些相当有争议的哲学命题。通常我们假定，当科学家们用一种新的理论来替代既存理论时，他们都是在客观证据的基础上这样做的。但是库恩认为，接受一种新的范式是科学家出于信念的一种特定行为。他承认，一个科学家可能会有很好的理由为一种新范式而放弃旧范式，但是他强调单靠这些理由永远无法合理地**迫使**范式转变。库恩写道，"从信奉一个范式到信奉另一个范式，这是一种不受强迫的转变经历"。在解释为什么一种新范式在科学共同体内能够快速获得认同的问题时，库恩重点强调了科学家们之间相互的同行压力。如果一种特定的范式拥有强有力的倡导者，它就更有可能赢得广泛的认同。

库恩的许多批评者都对这些主张感到震惊。如果范式转换是以库恩所说的那种方式实现的，就真的难以理解科学如何能被看做一种理性活动。科学家们真的不得不将信念建立在证据和理性，而不是信念和同行压力的基础上吗？面对两种竞争的范式，科学家们确实应该进行客观比较以决定哪种范式有着更多的有利论据吗？接受"改宗"，或甘愿被最强势的同行科学家说服，这些似乎很难算得上理性的行事方式。库恩对于范式转换的解释,似乎也很难与实证主义者眼中作为一种客观、理性活动的科学相容。一位评论者写道,按照库恩的解释,科学中的理论选择就是"群众心理学的事"。

库恩也对科学变迁的总体方向提出过有争议的看法。按照一种广泛接受的观点,科学进步总是以线性的方式趋近真理,旧的不正确的观点总是被新的正确的观念所取代。新近的理论因而在客观性上要优于早先的理论。科学的这一"累积性"概念在科学家和外行人中一样通行,但是库恩认为,这既是历史的不准确又是哲学的幼稚。例如,他指出爱因斯坦的相对论在某些方面与亚里士多德而不是牛顿的理论更为相似——所以力学的历史就不仅仅是一种从错误到正确的线性进步。此外,库恩还质疑客观真理的概念是否真正有意义。在他看来,认为存在着一系列确定的、独立于任何范式之外的关于世界的事实,这种想法的融贯性是值得怀疑的。库恩提出了一种激进的替代思想:关于世界的事实都是系于范式的,当范式变化时它们也要发生变化。如果这一主张正确,那么问一种特定的理论是否与"本然的"事实相关,或者因此询问这种特定的理论是否是客观真理,就都没有意义。

不可通约性和观察数据的理论负荷

关于上述论断库恩有两点主要的哲学论证。首先,他认为竞争的范式之间一般是互相"不可通约的"。要理解这一思想,我们必须记住,对库恩来说一个科学家的范式决定了她的总体世界观——她通过范式的透镜去看取一切。所以在科学革命中,当现存范式被新范式取代时,科学家们必须放弃他们用以了解世界的整个概念框架。事实上,库恩甚至明显带有比喻意味地声称,在范式转换的前与后科学家们"生活在不同的世界里"。不可通约性是指,两个范式间是如此不同以至于不可能对两者进行任何直接的比较——没有一种共同的语言实现互通。于是,库恩说,不同范式的支持者们"不能充分交流彼此的观点"。

这即使有些含混,却是一个有趣的想法。不可通约性的说法主要来自库恩的信念,即科学概念的意义来自用到这些概念的理论。所以,(例如)要理解牛顿的质量概念,就需要理解牛顿物理学的整个理论——概念不能独立于它们所嵌入的理论而获得解释。这一有时被称为"整体论"的思想,受到库恩的特别重视。他认为,"质量"这一术语对于牛顿和爱因斯坦来说实际上所指不同,因为都包含这一术语的两种理论是如此地不同。这意味着牛顿和爱因斯坦实际上是在说着不同的语言,显然这使得在两种理论间作出选择的努力变得复杂。如果一位信奉牛顿学说的物理学家和一位信奉爱因斯坦学说的物理学家尝试着进行一次理性讨论,那么他们的谈话将以缺乏交集而告终。

库恩使用不可通约性命题既是为了反驳范式转换是完全"客观的"这一观点,也是为了支持自己提出的非累积性的科学史图景。传统的科学哲学在两个竞争的理论中进行选择时没有什么大的困难——仅需借助有效证据就可以对两种理论作出客观比较,进而决定哪种更好。但是,这明显假设了存在着一种可以表述这两种理论的共同的语言。如果库恩是正确的,即新旧范式的支持者们只是在相当字面的意义上缺乏交集,对于范式选择的这种简单化解释就不可能正确。对于传统的"线性"科学史图景来说,不可通约性也同样成问题。如果新旧范式之间不可通约,把科学革命看做是"正确"思想取代"错误"思想也就不对。声称一种思想是正确的而另一种是错误的,就暗示着存在一种评价它们的共同框架,而这一点正是库恩所否定的。不可通约性暗示着科学变迁在某种意义上是无方向的,远非一种朝向真理的直线式进步:新近的范式并不比先前的更好,仅仅是不同罢了。

库恩的不可通约性命题并没有令许多哲学家信服。部分问题在于,库恩也承认新旧范式是**不相容的**。这一主张看起来很有道理,因为如果新旧范式并非不相容,就没有必要在两者之间选择。在许多例子中,这种不相容性也是显而易见的——托勒密的行星围绕着地球运转的主张显然与哥白尼的行星围绕太阳运转的主张不相容。但是正如库恩的批评者们立即指出的,如果两种事物不可通约,它们就不可能是不相容的。为了理解原因,我们来看一个命题:物体的质量取决于速度。爱因斯坦的理论认为这一命题正确,而牛顿学说认为这一命题错误。但如果不可通约性学说正确,牛顿和爱因斯坦在此就不存在真正的分歧,因为这一命题在两种理论中意思不同。仅当这一命题在两种理论中具有**同样的**意义,也即仅当不存在不可通约性时,两者之间才有真正的冲突。既然每一个人(包括库恩)都承认爱因斯坦和牛顿的理论的确有冲突,我们就有充分的理由质疑不可通约性命题。

为了回应这类反驳,库恩稍微缓和了他的不可通约性命题。他强调即使两种范式不可通约,也不意味着两者之间无法比较,而仅仅是使得这种比较更为困难。库恩认为,不同范式间的**部分**转化仍是可能的,所以新旧范式的支持者们在某种程度上可以相互交流:他们不会总是完全缺乏交集。但是库恩又进而主张,两种范式间完全客观的选择是不可能的。因为除了由于缺乏共同语言而导致的不可通约性外，还存在着库恩所称的"标准的不可通约性"。这一概念是指，不同范式的支持者们在关于评价范式的标准，关于一个好的范式应该解决什么问题，关于那些问题可接受的解决办法如何等方面意见可能不一致。所以即便能够有效交流，他们在哪种范式更好这一问题上也不可能达成一致。用库恩的话说，"每一种范式都能满足它为自己所设定的标准，却达不到它的反对者们所设定的一些标准"。

库恩的第二个哲学论证建立在数据的"理论负荷"这一观念上。为了理解这一观念，假设你是一个试图在两种竞争的理论间作出选择的科学家。显然要做的事情是，去寻找有助于在两种理论间进行取舍的一系列观察数据——这正是传统科学哲学所推崇的。但这种做法仅在存在着适当独立于理论的观察数据时才可能，也就是说，无论相信两种理论中的哪一种，科学家都会接受观察数据。正如我们所看到的，逻辑实证主义者认为存在着这种理论中立的观察数据，它们能够为两种竞争理论提供客观的裁决。但是库恩认为，理论中立概念是一种幻相——观察数据总会受到理论假设的感染。将所有科学家都认为与各自理论流派无关的一组"纯粹"数据隔离出来是不可能的。

对于库恩来说，观察数据的理论负荷具有两项重要意义。首先，它意味着竞争范式之间的问题不能通过简单诉诸"数据"或"事实"来解决，因为被科学家称为数据或事实的东西依赖于她所接受的范式。在两种范式间作出完全客观的选择因此是不可能的：不存在能够评价各方主张的中立见解。第二，客观真理的概念本身也值得质疑。要达到客观真理，我们的理论或信念必须与事实相符，但是如果事实本身就受到理论影响，这种相符的观念就变得没有意义了。这就是为什么库恩倒向了一种激进的观点：真理本身是相对于范式而言的。

为什么库恩会认为所有的观察数据都是理论负荷的？他的著作中并没有彻底澄清这一点，但从中至少可以看出两条论证思路。第一条思路是，知觉很大程度上受制于背景信念——我们所看到的部分取决于我们所相信的。一个训练有素的正在实验室里看着精密仪器的科学家所看到的与一个门外汉所看到的将会有所不同，因为科学家显然具有门外汉所没有的关于仪器的许多信念。大量心理学实验据称表明了知觉以这一方式对背景信念敏感——尽管对这些实验的解释仍有争议。第二条思路是，科学家的实验报告和观察报告通常是用高度理论化的语言来表述的。例如，一位科学家可能以"一束电流正通过一根铜棒"来描述实验结果。但是这一数据描述显然负荷了大量理论。一位不对电流持标准信念的科学家就不会接受它，因此它明显不是理论中立的。

在上述论证的价值上，哲学家们发生了分歧。一方面，许多哲学家赞同库恩，认为纯粹的理论中立是一种无法达到的理想。实证主义者的观念，即完全没有理论倾向的一组数据陈述，已被大多数当代哲学家所否定——部分也是因为，没有人曾成功地具体说明这种陈述。然而，这是否完全损害了范式转换的客观性，这一点并不清楚。例如，假设托勒密主义的天文学家和哥白尼主义的天文学家相互争论哪种理论更好。争论要有意义，就一定要存在着一些他们都能赞同的天文学数据。但为什么这会是一个问题？他们对（例如）地球和月亮在前后相继的夜晚所处的相对位置或者太阳升起的时间等问题必定有一致看法吗？显然，如果哥白尼主义者坚持用日心说描述观察数据，托勒密主义者将会加以反对。然而哥白尼主义者没有理由那样做。类似"5月14日太阳在早上7点10分升起"这样的陈述能被任何一方科学家接受,无论他们相信地心说还是日心说。这样的陈述可能不是**完全**理论中立的,但它们已充分摆脱了理论影响以至于两种范式的支持者们都能接受,这才是最重要的。

　　数据的理论负荷迫使我们放弃客观真理的概念,这一点更不够明显。许多哲学家都承认,理论负荷使人难以看出对客观真理的**了解**何以可能,但是这并不意味着客观真理这一概念本身是不自洽的。问题部分在于,与许多怀疑此概念的人一样,库恩也不能清晰提出一个可行的替代方案。真理是相对于范式而言的——这种激进观点终究难以理解。因为,与所有这类相对主义学说一样,它也面临着一个关键的问题。看看下面的问题:真理是相对于范式而言的这一论断**自身**在客观上是对是错?如果相对主义的支持者回答"对",他们就已经承认了客观真理的概念有意义,因而也自相矛盾了。如果他们回答"错",他们就没有理由反驳与他们意见不一并声称真理**并非**相对于范式而言的人。并不是所有的哲学家都认为该论证对于相对主义来说完全致命,但它的确表明,放弃客观真理的概念说易行难。传统观点认为科学史仅仅是一种朝向真理的线性进步,库恩的确对这一观点提出了一些有力的反对意见,但是他提出的相对主义替代方案也存在着诸多问题。

库恩和科学的合理性

　　《科学革命的结构》一书行文非常激进。库恩处处给人以这样的印象:要以一种全新的观念取代关于科学领域理论变化的标准哲学观念。他的范式转换学说、不可通约性学说以及观察数据的理论负荷学说，似乎与实证主义者把科学看做一种理性、客观、累积的事业的观念格格不入。大多数库恩的早期读者有充分理由认为，他是在声称科学是一种完全与理性无关的活动，其特征是在常规时期教条地坚持一种范式，在革命时期突然"改宗"。

但是库恩自己对于这种解读并不高兴。在 1970 年出版的《科学革命的结构》第二版的后记以及后来的论著中，库恩在很大程度上缓和了他的激进论调——并且指责某些早期读者误解了他的意图。库恩声称，他的书并不是要怀疑科学的合理性，而是要提供一种关于科学事实上如何演变的更为实在、更符合历史的图景。由于无视科学史，实证主义者滑入了对科学活动的过分简单化甚至理想化的解释，而库恩的目的就在于进行纠正。他并不试图表明科学是非理性的，而是要提供一种对科学之合理性的更好解释。

一些评论者把库恩的后记仅看成一种转变——一种从原先立场的退却，而不是对原先立场的澄清。这是否是一种公平的评价不是我们接下来要讨论的。但是后记的确揭示出一个重要方面。在对那些指责他将范式转换描绘为无关理性的人进行反驳时，库恩提出了一个著名的观点：科学领域的理论选择"没有算法"。这是什么意思呢？一种算法就是指一系列使我们能计算出关于特定问题的答案的规则。例如，乘法的算法就是一种把它运用到任何两个数字上就能得出乘积的一套规则。（当你在小学学习算术的时候，其实就是在学习加、减、乘、除的算法。）因此理论选择的算法就是指一系列的规则，当被运用到两个竞争的理论中时它们能告诉我们应该选择哪一理论。实际上大多数实证主义的科学哲学都依赖于这样一种算法的存在。实证主义者们似乎经常写道,只要给定一系列观察数据和两个竞争的理论,我们就能使用"科学方法的原则"去决定哪一种理论更优。这一思想隐含于实证主义者的如下信念:虽然发现是心理学的事,证明却是逻辑的事。

库恩坚持认为科学领域的理论选择不存在算法,这一点几乎肯定是正确的。还没有人曾成功发明这样一种算法。许多哲学家和科学家对于要在理论中寻求什么这一问题提出过貌似合理的意见——简单性、适用范围的广泛、与数据的契合,等等。但正如库恩所深知的,这些意见还远远达不到提出一种真正的算法。首先,理论之间可能会有些权衡:理论1可能比理论2更简单,但是理论2可能更符合观察数据。所以主观判断或科学常识在对两个竞争理论的裁定中常常要用到。从这一角度看,库恩关于新范式的采用涉及特定信念行为的观点似乎并不特别激进,同样,他所强调的,即在决定一种范式在科学共同体中的胜出几率时其倡导者的说服作用,也不是非常极端。

理论选择不存在算法这一论题支持了另一观点,即库恩对于范式转换的解释没有攻击科学的合理性。因为,我们可以把库恩解读成是在拒斥一种特定的合理性概念。事实上实证主义者认为,理论选择**一定**存在着一种算法规则,否则科学变迁就是非理性的。这绝不是一种疯狂的观点:许多理性行动的范例都涉及到规则,或者说算法。例如,如果你想知道某种商品是在英国还是在日本便宜,你就会运用一种算法将英镑换算成日元;任何其他寻求答案的方法都是非理性的。类似的,如果一位科学家正试图在两个竞争的理论间作出决定,人们很容易认为唯一理性的方法是运用一种关于理论选择的算法。因此,如果已经表明没有这样一种算法规则(看来很可能如此),我们就面临两种选择。要么我们能得出科学变迁非理性的结论,要么就是实证主义者的合理性概念太苛求了。在后记中库恩表示,后一种选择是对他的著作的正确解读。他的书的寓意并不在于范式转换是非理性的,而是要表明,要理解范式转换,我们需要一种更为宽松的、非算法的合理性概念。

库恩的遗产

虽然库恩的思想存有争议,但是它们改变了科学哲学。这部分是因为库恩质疑了许多传统上被视为理所当然的假定,迫使哲学家们重新正视它们,部分是因为他引起了对传统科学哲学完全忽视的一系列问题的关注。库恩之后,认为哲学家们能够忽视科学史的观念越来越站不住脚,正如在发现和证明的语境间作出截然区分的观念日益站不住脚一样。当代科学哲学家比前库恩时期的前辈们远为关注科学的历史演变。即使是那些对库恩的激进观点持反对意见的人,也承认在这些方面库恩的影响是积极的。

库恩的另一个重要影响是使我们的关注聚焦在科学发生的社会情境中,这一点是传统科学哲学所忽视的。对于库恩来说科学本质上就是社会活动:一个通过遵从一种共有范式联结起来的科学共同体的存在是常规科学实践的先决条件。库恩还对在学校和大学里如何教授科学、年轻科学家们如何被吸纳到科学共同体中、科学成果如何发表等其他类似的"社会学"问题投入了大量关注。无怪乎库恩的思想在科学社会学家中间具有深刻影响。尤其是,库恩对科学社会学领域 20 世纪 70 年代始于英国、被称为"强纲领"的运动功不可没。

强纲领立基于如下思想:科学应该被视为作为科学实践之

场所的社会的产物。强纲领的社会学家们认真地对待这一思想：他们认为科学家的信念大部分是由社会决定的。所以，(例如)要解释为什么一位科学家相信某一特定理论，他们会去引证该科学家所处的社会文化背景。他们坚称，科学家相信这一理论的个人原因从来都不是充分的解释。强纲领借用了库恩的大量命题，包括数据的理论负荷、科学本质上是一种社会事业，以及理论选择没有算法。然而强纲领社会学家们比库恩更为激进，却没有库恩那样谨慎。他们公开否认客观真理和合理性的概念，认为这些概念在观念形态上可疑，并且对传统的科学哲学抱有极大的怀疑。这使得科学哲学家和科学社会学家之间产生了一些紧张，一直持续到今天。

说得更远点，库恩还对**文化相对主义**在人文社会科学中的兴起产生了影响。文化相对主义不是一种有着精确定义的学说，但其核心思想是不存在类似绝对真理的东西——真理总是相对于特定的文化而言的。我们可能认为西方科学揭示了关于世界的真理，但是文化相对主义者们可能会说其他的文化和社会，例如土著美国人，拥有他们自己的真理。正如我们看到的，库恩确实信奉相对主义的观点。然而，在他对文化相对主义的影响方面事实上存在着一种反讽。文化相对主义者们通常是非常反科学的。他们反对社会赋予科学的崇高地位，认为这是对其他具有同等价值的信念系统的歧视。但库恩本人是坚决支持科学的。与实证主义者一样，他认为现代科学是一种有着深远影响的智识成就。库恩的范式转换、常规科学和革命科学、不可通约性和理论负荷等学说并不是有意破坏或批判科学事业，而是要帮助我们更好地理解科学。

物理学、生物学和心理学中的哲学问题

迄今为止我们所探讨的问题——归纳、解释、实在论和科学变迁——都属于所谓的"一般科学哲学"。这些问题都是关注一般意义上的科学探究的本质,而不是特别地与(例如)化学或地质学相关的本质。然而,特定科学中也有许多有趣的哲学问题,这些问题属于我们所说的"特殊科学的哲学"。这些问题通常部分依赖于哲学沉思,部分依赖于经验事实,从而十分有趣。在本章中,我们要考察分别来自物理学、生物学和心理学的三个这类问题。

莱布尼兹 VS.牛顿:关于绝对空间

我们的第一个主题是关于 17 世纪两个杰出科学家——莱布尼兹(1646—1716)和牛顿(1642—1727)之间就时空本质的争论。我们将主要关注空间问题,而时间问题也与此紧密相关。在其著名的《自然哲学原理》一书中,牛顿为一种被称为"绝对主义者"的空间观念进行了辩护。根据这种观点,空间拥有一种"绝对"存在,超越于各种物体的空间关系之上。牛顿把空间看做一个三维的容器,上帝在创造世界的时候把物质世界放在其中。这就意味着空间在有物体之前就已存在,正如在把食品放进食品盒之前该容器就已经存在一样。根据牛顿的看法,空间与食品盒之类的日常容器之间的区别仅在于,日常的容器显然尺寸有限,而空间却在每个方向上都无限延伸。

莱布尼兹强烈地反对这种绝对主义的空间观和牛顿哲学中的许多其他观点。他认为空间仅是由物体间的空间关系构成的集合。"上""下""左""右"就是空间关系的个例——它们是物体相互之间具有的关系。这种"关系论者"的空间概念意味着,在有物体之前空间并不存在。莱布尼兹认为在上帝创造物质世界**之时**,空间才开始存在;空间并不是预先存在着并等待物体填充进去。所以把空间设想成一个容器甚或任何种类的实体都不是有效的想法。可以通过一个类比来理解莱布尼兹的观点。一份合法的合同由两方——如一所房子的买方和卖方——之间的关系构成。如果其中一方去世,合同便终止。所以,说合同独立于买卖双方间的关系而存在是不切实际的——合同就**是**这种关系。同样,空间也不是什么超越于物体间的空间关系而存在的东西。

牛顿引入绝对空间的概念主要是为了区别绝对运动和相对运动。相对运动是一个物体相对于另一物体的运动。就相对运动而言,问一个物体是否"真正"在运动是没有意义的——我们只能问它是不是相对于另一物体在运动。想象一下,两人沿着一条直道一前一后慢跑着,相对于站在路边的旁观者,这两个人明显处在运动中:他们正离得越来越远。但是这两个慢跑者相对于彼此却没有运动:只要他们保持同样的速度跑向同一方向,他们的相对位置就仍然不变。所以一个物体相对于一物可能处于运动之中,而相对于另一物却处于静止状态。

牛顿相信,绝对运动同相对运动一样也是存在的。常识支持这种观点。直观上,问一个物体是否"真正地"在运动**确实**是有意义的。想象处于相对运动中的两个物体——如一架空中的滑翔机和地面上的一位观察者。现在相对运动是对称的:正如滑翔机相对于地面上的观察者是运动的,地面上的观察者相对于滑翔机也是运动的。但是问以下问题是否确实有意义:观察者或滑翔机,或者两者,是否"真正地"在运动? 如果确有意义,我们就需要绝对运动的概念。

绝对运动到底**是**什么? 在牛顿看来,它是**相对于绝对空间自身**的物体运动。牛顿认为在任何时间,每个物体都在绝对空间中有一个特定的位置。如果一个物体从一个时刻到另一个时刻在绝对空间中改变了位置,该物体就处于绝对运动状态;反之,则处于绝对静止状态。所以为区分相对运动和绝对运动,我们需要把空间看做一个绝对的实体,超越于物体之间的关系。请注意,牛顿的推理依赖于一个重要的假设。他毫无疑问地假设所有运动都是相对于某个参照物的。相对运动是相对于其他物体的运动;绝对运动是相对于绝对空间自身的运动。所以在某种意义上,对于牛顿来说,即使绝对运动也是"相对的"。实际上牛顿是在假设,处于运动状态,不论绝对运动还是相对运动,不可能是关于物体的"原初事实";它只能是关于物体与其他事物间关系的事实。这里的其他事物可能是另一个物体,也可能是绝对空间。

莱布尼兹承认,在相对运动和绝对运动之间存在着区别,但是他反对把绝对运动解释为与绝对空间相关的运动。他认为绝对空间的概念是不严密的。对此,他作了大量论证,其中许多在本质上是神学的。从哲学的观点看,莱布尼兹最有趣的论证是,绝对空间与他所说的不可区分事物的同一性原则(PII)相矛盾。莱布尼兹认为这条原则毋庸置疑是正确的,所以他拒斥绝对空间的概念。

不可区分事物的同一性原则指的是,如果两个物体不可区分,它们就是同一的,即它们实际完全是同一个物体。说两个物体不可区分意味什么呢?这意味着根本不能在这两者之间找到任何区别——两者具有完全相同的属性。所以如果该原则是真的,那么任何两个真正不同的对象必须至少在一个属性上不同——否则它们就是同一个而不是两个物体。不可区分事物的同一性原则在直观上非常具有说服力。找到两个不同的物体共同具有**所有**属性的例子当然不容易。甚至工厂里大批量制造的两个产品通常也会在许多方面不同,即便这些差别不能通过肉眼观察到。该原则总体上是否正确,是哲学家们仍在争论的复杂问题;答案部分取决于究竟什么能被算做"属性",部分取决于量子物理学中的疑难问题。但是我们目前关注的是莱布尼兹对这条原则的应用。

莱布尼兹用了两个思想试验来揭示牛顿的绝对空间理论和不可区分事物的同一性原则之间的矛盾。他的论证策略是间接的:为了论证,他假设牛顿的理论正确,然后他力图证明这一假设会带来矛盾;矛盾不可能为真,所以莱布尼兹的结论是牛顿的理论必为假。回想一下牛顿的观点,他认为在时间上的任何时点,宇宙中的每一个物体在绝对空间中都有一个确定的位置。莱布尼兹要我们设想两个不同的宇宙,其中包含有彼此完全相同的物体。在宇宙 1 中,每个物体在绝对空间中都占据一个特定的位置。在宇宙 2 中,每个物体在绝对空间中都被移到了一个不同的位置,(例如)向东移了两英里。没有任何方式可以区分开这两个宇宙。因为正如牛顿自己所承认的,我们不能观察到绝对空间中物体的位置。我们所能观察到的只是物体**相对于其他物体**的位置,而这些位置没有变化——所有物体在移动的量上都相同。任何观察和实验都永远不能揭示出我们是生活在宇宙 1 还是宇宙 2 中。

第二个思想试验与第一个相类似。回想一下牛顿的理论,他认为一些物体在绝对空间中移动而另外一些物体处于静止状态。这就意味着在每一时刻,每个物体都有一个确定的绝对速度。[速度(velocity)是在一定方向上的速度(speed),所以一个物体的绝对速度是该物体在绝对空间中一定方向上的移动速度。绝对静止的物体绝对速度为零。]现在想象有两个不同的宇宙,它们之中有完全相同的物体。在宇宙 1 中,每个物体都有一个特定的绝对速度。在宇宙 2 中,每个物体的绝对速度都增加了一个固定的量,比如说(增量为)在一个规定的方向上每小时 300 公里。我们还是永远不能区分开这两个宇宙。因为正如牛顿自己所承认的,我们不可能观察到一个物体相对于绝对空间的移动速度。我们只能观察到物体**相对于其他物体来说**移动的速度——而这些相对速度将保持不变,因为每个物体的速度都增加了完全相同的量。没有任何观察和实验能够揭示出我们是生活在宇宙 1 还是宇宙 2。

在上述每个思想试验里,莱布尼兹都描绘了两个宇宙,用牛顿自己的理论永远无法区分开——它们完全不可分辨。但是根据不可区分事物的同一性原则,这就意味着这两个宇宙实际上是同一个。所以结果就是,牛顿的绝对空间理论是错误的。还可以用另外一种方式来看待这一点:牛顿的理论暗示,处于绝对空间中某一处的宇宙与移动到不同处的该宇宙之间有真正的差异。但是莱布尼兹指出,只要宇宙中每个物体位置移动的量相同,这种差异就完全不可觉察。如果在两个宇宙之间觉察不到任何差异,它们就是不可区分的,不可区分事物的同一性原则告诉我们,这两个宇宙实际上是同一个。所以牛顿理论的一个错误是:它在只有一个事物的时候认为有两个事物存在。绝对空间的概念因而与不可区分事物的同一性原则相冲突。莱布尼兹的第二个思想试验逻辑与此相同。

　　实际上,莱布尼兹是在声称绝对空间是一个空概念,因为它不能在观察上作出区分。如果既不能觉察到绝对空间中物体的位置,也不能觉察到相对于绝对空间的物体速度,那又为什么要相信绝对空间?莱布尼兹在此诉诸一个非常合理的原则,即仅当不可观察的实体的存在会带来能够观察到的差异,我们才应该在科学中假定该实体存在。

　　但是牛顿认为他能够揭示绝对空间**确实**有可观察的效应。这就是他著名的"旋转桶"论证的要点所在。他让我们想象一个装满了水的桶,它由一根穿过位于其底部的一个孔洞的绳子悬挂着(见图 12)。

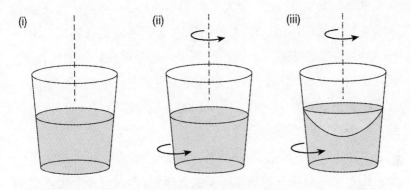

图 12　牛顿的"旋转桶"试验。在步骤 i,桶和水都静止;在步骤 ii,桶相对于水在转;在步骤 iii,桶和水协力地转动。

最初水相对于桶处于静止状态,然后绳子被搓动了许多次再放开。随着绳子的展开,桶开始旋转。起先,桶中的水保持静止,水面是平的;桶相对于水在旋转。但是稍后,桶把它的运动传递给水,水也开始随着桶协力旋转;桶与水相对于彼此又静止了。操作显示,之后水面如图所示在桶边处向上凸起。

是什么造成水面的隆起? 牛顿问道。明显这与水的旋转有关。但旋转是运动的一种类型,而对牛顿来说,物体的运动总是相对于其他物体的。所以我们必然要问:水相对于什么在旋转? 显然不是相对于桶,因为桶和水在协力旋转,因而它们之间相对静止。牛顿认为水是相对于绝对空间在旋转,并且这导致了水面的向上凸起。所以绝对空间的确在事实上有可观察的效应。

你也许会认为牛顿的论证中有个明显的缺陷。就算水不是相对于桶在旋转,为什么就能得出一定是相对于绝对空间在旋转?水的旋转是相对于做这个实验的人,相对于地球的表面,以及相对于固定的星辰,是否其中的任何一个当然都有可能导致水面的隆起? 牛顿对这一运动有个简单的回答。想象一个只包含该旋转的桶的宇宙。在这个宇宙中,我们不能用水是相对于其他物体在旋转来解释水面的隆起,因为不存在其他物体,并且与之前一样,水相对于桶是静止的。绝对空间是剩下的水的旋转唯一可以相对的东西。所以我们必须相信绝对空间,不然就不能解释为什么水面会隆起。

实际上,牛顿是在说,尽管一个物体在绝对空间中的位置和它相对于绝对空间的速度不能被觉察到,但说出一个物体相对于绝对空间何时在**加速**却**的确**可能。因为当一个物体旋转时,根据定义它就在加速,即使旋转的速率不变。这是因为在物理学上,加速度被定义成速度变化的比率,并且速度是**一定方向上的**速度。旋转的物体一直在改变着运动的方向,结果就是它们的速度不是不变的,因此它们在加速。隆起的水面恰恰就是所谓"惯性效应"——由加速运动产生的效应——的一个例子。另外一个例子是当飞机起飞时,你所获得的被推向椅背的感觉。牛顿坚信,惯性效应的唯一可能的解释是,经受那些效应的物体相对于绝对空间在加速。在一个只有加速物体的宇宙中,绝对空间是加速度唯一能够相对的。

牛顿的论证很有力,但不能说服人。因为如果旋转桶试验是在一个没有其他物体的宇宙中完成的,牛顿如何知道水面**会**向上隆起?牛顿想当然地假设,我们在这个世界中所发现的惯性效应在没有其他物体的世界中也会保持不变。这明显是个非常重要的假设,许多人已经质疑牛顿有何理由如此设想。所以,牛顿的假设不能证明绝对空间的存在。相反,它为莱布尼兹的辩护者平息了来自外界的一种挑战,即要求他们提出惯性效应之外的替代解释。

莱布尼兹也面临着不借助绝对空间来解释绝对运动和相对运动之间区别的挑战。在这个问题上,莱布尼兹撰文称,"当实体变化的直接原因在实体本身时",该实体就在真正地或绝对地运动。回想一下滑翔机和地面上的观察者的例子,相对于彼此,两者都在运动。为了确定哪个在"真正地"运动,莱布尼兹会说我们需要确定变化(即相对运动)的直接原因是在滑翔机、观察者还是这两者。这种关于如何区别绝对运动和相对运动的提议,避免了一切对绝对空间的参照,但是却很不清晰。莱布尼兹从未严格地解释过在一个物体中"变化的直接原因"是什么**意思**。但是也许他的意图是要拒斥牛顿的假设,即一个物体的运动,不管是相对运动还是绝对运动,都只能是关于该物体与其他物体间关系的一个事实。

令人感兴趣的是,关于绝对和相对的争论并没有消逝。牛顿关于空间的论述与他的物理学有密切的关系,而莱布尼兹的观点是对牛顿观点的直接回应。所以也许有人会认为17世纪以来的物理学的发展,到目前应该已经解决了这一问题。但是这却没有发生。尽管人们曾经普遍认为,爱因斯坦的相对论已经作出了偏向于莱布尼兹的论断,但是近些年来这种观点日益遭到批判。源于牛顿和莱布尼兹之间的争论在300多年后变得更为激烈。

生物学分类的问题

分类,或者说把正在研究的对象归到一般的种类中,在每门科学中都起到作用。地理学家按形成方式把岩石分为火成岩、沉积岩以及变质岩。经济学家按公平程度将税制分为比例税制、累进税制及累退税制。分类的主要作用是传达信息。如果化学家告诉你某物是金属,那就告诉了你很多关于它的可能性状。分类提出了一些有趣的哲学问题。这些问题大部分源于这一事实,即任何给定的对象集合原则上都可以按很多不同的方式来划分类别。化学家根据物质的原子数目来划分物质,产生了元素周期表。但是他们同样也能按照物质的颜色、气味或密度来划分物质的类别。我们该如何在这些可能的分类方式中作出选择?存在一种"正确的"分类方式吗?或者是否所有的分类方案最终都是任意的?这些问题在生物分类或分类学中显得特别紧要,正是我们在此要关注的。

生物学家传统上用林奈系统来划分植物和有机生物,这一系统是以 18 世纪瑞典博物学家卡尔·林奈（1707—1778）来命名的（见图 13）。林奈系统的基本元素对许多人来说简单而熟悉。首先，个体的有机生物属于一个**种**，然后每个物种属于一个**属**，每个属又属于一个**科**，每个科属于一个**目**，每个目属于一个**纲**，每个纲属于一个**门**，而每个门又属于一个**界**。多种中间等级，如**亚种、亚科**和**总科**也被加以识别。种是基本的分类单元，属、科、目等等被视做"高级分类单元"。一个物种的标准拉丁名指示了该物种所归入的属，仅此而已。例如，你和我都属于**智人**，这是人属中唯一存活的物种。人属中其他两个物种是**直立人**和**能人**，这两个种现在都已经灭绝了。人属又属于人科，人科又属于类人猿总科，类人猿总科又属于灵长目，灵长目又属于哺乳动物纲，哺乳动物纲又属于脊索动物门，脊索动物门属于动物界。

　　应该注意，林奈划分有机生物的方法是层级式的：众多种处于单个属中，众多属又处于单个科中，众多科又处于单个目中，以此类推。所以当我们向上推移时，会发现每个层上的分类单元越来越少。在底部差不多有数百万物种，但是到了顶部仅有五个界：动物、植物、真菌、细菌和原生物（海藻、海草等）。并非科学中的每个分类系统都是等级式的。化学中的周期表就是非等级式分类的一个例子。不同的化学元素并不是像林奈系统中种的划分方式，被安置在越来越具有总括性的分组中。我们必须面对的一个重要问题是，生物学分类**为什么**应该是层级式的。

　　林奈系统在过去几百年中一直很好地满足了博物学家们的需要，并且一直被延用至今。在某些方面这令人惊讶，因为在这段时期内生物学理论已经发生了很大改变。现代生物学的奠基石是达尔文的进化论，这一理论认为当代的物种源自远祖物

CAROLI LINNÆI

Naturæ Curioforum *Diofcoridis Secundi*

SYSTEMA NATURÆ

IN QUO

NATURÆ REGNA TRIA,

SECUNDUM.

CLASSES, ORDINES, GENERA, SPECIES,

SYSTEMATICE PROPONUNTUR,

Editio Secunda, Auctior.

STOCKHOLMIÆ

Apud GOTTFR. KIESEWETTER,

1740.

图 13　林奈的名著《自然系统》，该书介绍了他对植物、动物和矿物质的分类。

种;这种理论与古老的、圣经所启示的观点相冲突,后者认为每一物种都是被上帝独立创造出来的。达尔文的《物种起源》一书于 1859 年出版,但是直到 20 世纪中叶,生物学家才开始发问进化论是否应该影响有机体分类的方式。直到 20 世纪 70 年代,两个对立的分类学派才出现,这两个学派为该问题提供了竞争性的解答。按照**分支分类学派**的观点,生物学分类应该力图反映物种间的进化关系,所以进化史的知识对于作出好的分类是不可或缺的。但根据**表现型分类学派**的观点,情况却不是这样:分类学能够而且应该完全独立于进化方面的考虑因素。第三个派别被称做**进化分类学派**,他们力图把前两者的观点结合起来。

为了理解分支分类学派和表现型分类学派之间的争论,我们必须把生物学分类的问题一分为二。第一个问题是如何把有机体划归到种中去,这被称做"物种问题"。这个问题远没有得到解决,但是在实际中生物学家通常能够就如何划定物种的界限达成一致,尽管也有一些很难划界的情况。一般而言,如果有机体相互之间能够杂交繁殖,生物学家就把这些有机体归为同一种,反之,就把它们归为不同种。第二个问题是把一组物种归入到更高级的分类单元中去,这显然预设第一个问题已经有了解决方案。正如所发生的那样,分支分类学派和表现型分类学派虽然通常在物种问题上不能达成一致,但是他们之间的争论主要集中在更高级的分类单元上。所以此刻,我们先忽略物种问题——假设有机体已经以一种令人满意的方式被归入所属的种当中去了。问题是:下一步该怎么办?我们要使用什么原则来把这些种划分到更高级的分类单元中去?

为了突出这一问题,我们先来思考下面的例子。人类、黑猩猩、大猩猩、倭黑猩猩、猩猩和长臂猿通常被一起归入类人猿总科。但是狒狒又不算做类人猿,为什么会这样呢?把人类、黑猩猩和大猩猩等放在一组,而又不把狒狒放在该组中,理由是什么呢? 表现型分类学派的答案是,前一组都共有很多狒狒所没有的特征,例如没有尾巴。按照这种观点,分类学的编组应该基于**相似性**——应该把在重要方面相互类似的物种放在一起,排除不相类似的物种。直观上,这是一种合理的观点。因为它与分类的目的在于传达信息的观念是完全吻合的。如果分类学的分组基于相似性,知道一个特定的有机体属于哪个组就会告诉你很多关于它的可能特征。如果被告知一个给定的有机体属于类人猿总科,你将会知道它没有尾巴。而且,被传统分类学认可的许多分组似乎确实基于相似性。举个明显的例子,植物都具有动物所没有的很多特征, 所以从表现型分类学派的观点看,把所有植物放在一个界而把所有动物放在另一界是很合理的。

然而,分支分类学派坚称,分类不该考虑相似性。真正需要的是物种间的进化关系——我们所知的**种系发生**关系。分支分类学派同意狒狒应处于包括了人类、黑猩猩和大猩猩等的类群之外,但是如是判定的理由与物种间的相似和差异无关。真正原因在于,类人猿总科物种之间比它们与狒狒之间关联得更为密切。确切说来,这是什么意思呢?它意味着所有类人猿总科物种都有一个共同的祖先,这一祖先却不是狒狒的。需要注意的是, 这并**不是**说类人猿总科物种和狒狒根本没有过共同的祖先。相反,如果在进化的时间上你能够追溯得足够久远,任何两个物种都有一个共同的祖先——地球上的所有生命被认定为有唯一的起源。关键之处在于,类人猿物种与狒狒的共同祖先也是许多其他物种,如各种各样的猕猴种的祖先。所以分支分类学派声称,包括了类人猿总科物种和狒狒的任何分类学类群必须也包括这些其他的物种。任何分类学类群都不能够**仅仅**包括类人猿物种和狒狒。

分支分类学派的核心观点是,所有的分类群,不管是属、科、总科还是其他,都必须是**单系的**。单系类群包括一个祖先物种和所有它的后代,但是不包括其他任何物种。单系类群大小各不一样。在一极是所有曾经存在过的物种形成一个单源类群,假定地球上的生命只有过一次起源。在另一极是只有两个物种的单系类群——如果它们是一个共同的祖先仅有的后代。只包括类人猿总科物种和狒狒在内的类群不是单系的,因为正如我们所看到的,类人猿总科物种和狒狒的共同祖先也是猕猴的祖先。所以按照分支分类学派的观点,它就不是一个真正的分类群。不管类群的成员有多么简单,只要该类群不是单系的,它就不允许出现在分支分类学的分类中。因为分支分类学派认为,与"自然"的单系类群相比,这种分组完全是人为的。

单系的概念通过图形很容易理解。且看下图(图 14)——通常称为**进化树**,该图表示了六个同期的物种(A—F)间的种系发生关系。如果我们在时间上回溯得足够久远,这六个物种都有一个共同的祖先,但是一些物种比其他物种间联系得更为紧密。物种 E 和 F 有一个非常近的共同祖先——它们的分支在相当近的过去相交过。相反,物种 A 与其余的后代在很久之前就分道扬镳了。现在来看{D、E、F}这一组。它是一个单系类群,因为它包含且只包含了所有只属于一个(未命名的)祖先物种的后代,在节点"X",这一物种分为两枝。{C、D、E、F}组同样也是一个单系类群,{B、C、D、E、F}组也是一样。但{B、C、D、F}组却不是单系类群。原因在于,这四个物种的共同祖先也是物种

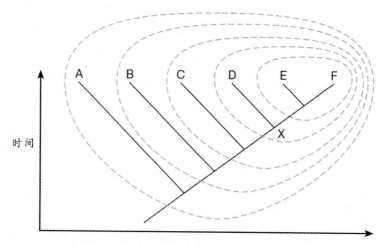

图 14 表示六个同期物种间的种系发生关系的进化树。

E 的祖先。图中所有单系类群都是环状的；任何其他的物种类群都不是单系的。

　　分支分类学派与表现型分类学派间的争论绝不纯粹是理论上的——有很多他们互有分歧的实际案例。一个著名的例子是关于爬虫纲或者爬行动物的。传统的林奈分类学认为蜥蜴和鳄鱼属于爬虫纲，但把鸟排除在爬虫纲之外而归入一个单独的鸟纲中。表现型分类学派赞成这一传统分类，因为鸟有其独特的、不同于蜥蜴、鳄鱼和其他爬行动物的身体结构和生理机能。但是分支分类学派主张爬虫纲根本不是一个真正的分类群，因为它不是单系的。如上图进化树（图 15）所示，蜥蜴和鳄鱼的共同祖先也是鸟的祖先；所以把蜥蜴和鳄鱼放在一个把鸟排除在外的类群中违背了单系性要求。分支分类学派因此建议放弃传统的分类习惯：生物学家根本不应该谈论爬虫纲，它是一个人造的而非自然的类群。这是一个非常极端的建议，即使那些赞同分支分类学精神的生物学家，通常都不愿意放弃被博物学家

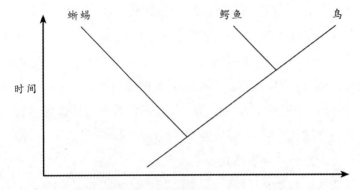

图 15 表示蜥蜴、鳄鱼和鸟之间的种系发生关系的进化树。

们很好地使用了几个世纪的传统分类范畴。

分支分类学派坚持认为,自己的分类方法是"客观的"而表现型分类学派的方法不是。这一指责当然有正确之处。因为表现型分类学派把物种间的相似性作为分类的基础,而对相似性的判断总会部分地含有主观成分。任何两个物种在一些方面都有相似之处,而在另一些方面不相似。例如,两个昆虫物种可能在身体结构上非常相似,但在摄食习惯上非常不同。那么,为了判断相似性,我们该选择哪些"方面"呢? 表现型分类学派希望通过定义一种"整体相似性"的标准来避免这一问题,这种标准将考虑一个物种的所有特征,这样就有可能建立起完全客观的分类。尽管这一想法听起来很好,但是它却不可行,主要是因为没有明显的计算特征的方法。当今很多人认为"整体相似性"的观念在哲学上是可疑的。表现型的分类确实存在,并用在了实践中,但是它们并非完全客观。对相似性的不同判断导致了不同的表现型分类, 没有明显的方法可用来在它们之间进行选择。

分支分类学派也面临着一系列它自己的问题。其中最严重的问题是,为了按照分支分类学的原则建立一个分类,我们就需要在设法分类的物种间弄清种系发生的关系,而这是非常困难的。仅通过观察这些物种显然不能弄清这些关系——它们只能通过推理得出。现在已经提出了多种推导种系发生关系的方法,但是它们还不十分完善。实际上,随着分子遗传学提出越来越多的证据,物种间种系发生关系的设想很快被推翻了。所以真正把分支分类学的思想变成实践是不容易的。在分类系统中只承认物种的单系类群当然省事,但是如果不知道一个给定的类群是否**是**单系的,这种方法用途就很有限。实质上,进化分类构建了关于物种间种系发生关系的假设,因而本来就是推测性的。表现型分类学派反对性地认为分类不应该在这方面有理论负荷。他们认为分类系统应该先于而非决定于对进化历史的推测。

　　尽管将分支分类学付诸实践存在着困难,并且分支分类学派在实际中常常建议对传统分类范畴进行相当根本性地修正,还是有越来越多的生物学家正转向这种分支分类学的观点。这主要是因为,分支分类学排除了表现型和其他分类法所具有的模糊性——它的分类原则尽管很难付诸实施,却非常清晰。并且,关于这一观点,即物种的单系类群是"自然的单元"而其他类群却不是,有一些非常直观的东西。此外,分支分类学还为生物学分类为什么应该是有层次的提供了真正的理由。如上面图15 所示,单系类群总是处在彼此的内部,如果严格遵循单系性要求,分类的结果就自然而然有层次。立足于相似性的分类方式也会引出层次性的分类,但表现型分类学家对于**为什么**生物学分类应该有层次却没有提供类似的解释。非常惊人的是,博物学家几百年前就已开始对有机生物进行层级式分类,但是如此分类的真正原因直到最近才弄清楚。

意识是模块化的吗？

心理学的一个主要工作是理解人类如何执行他们的认知任务。"认知任务"并不仅仅指解纵横字谜之类的事情，也指安全地过马路、理解他人所说的话、辨认别人的面容以及在商店里核对找零之类的普通任务。不能否认，人类非常擅长于其中的许多任务——如此擅长以至于我们通常做得很快，几乎不伴随有意识的思考。为了认识这一点有多么不寻常，让我们来考虑一个事实，即不管付出多大的努力和代价，机器人从来都没有被设计成哪怕只有一点点像人类在真实生活情境中那样行动。没有机器人能够像人类普通的一员那样机敏地解纵横字谜，或者参与一个对话。不知为何，人类能够最轻松地完成复杂的认知任务。我们所知的认知心理学，其主要解释目标就在于设法理解这是如何可能的。

我们所关注的焦点是在认知心理学家中由来已久且不曾间断的一个争论，它所涉及的是人类意识的建构。一种观点认为，人类意识是个"万能解题器"。这意味着意识中有一套通用的解题技巧，或"通用智能"，意识把它们运用于无限多的认知任务上。所以不管人们是在数弹子，决定去哪家饭馆吃饭，还是在努力学一门外语，所使用的都是同一套认知能力——这些认知任务代表了人类通用智能的不同应用。与此相对的另一种观点则认为，人类意识中包含大量专门的子系统或模块，每一种都是被设计用来执行非常有限的一类任务而不能执行其他任务（见图 16）。这被称做**意识的模块性**假说。例如，人们普遍相信有一个特殊的语言习得模块，这一观点源自语言学家诺姆·乔姆斯基。乔姆斯基认为,儿童并不是通过听取成人的谈话后用他们的"通用智能"来找出所说语言的规则;而是在人类儿童中有一种独特的、自行运转的"语言习得机制",它唯一的功能是,在适当刺激的情形下,让他或她学会语言。乔姆斯基为此论断提供了一系列给人深刻印象的证据——例如,甚至那些只有很低的"通用智能"的人通常也能通过学习把语言说得非常好。

模块性假说的一些最有说服力的证据来自于对脑损伤病人的研究,这种研究也被称为"缺陷研究"。如果人类心灵是万能解题器,我们就能预知,脑损伤会大致同等地影响所有认知能力。现实却并非如此。相反,脑损伤通常削弱某些认知能力而不伤及其他认知能力。例如,被称为"韦尼克区"的脑部的伤害会使得病人不能理解言语,尽管他们仍然能够说出流畅的、符合语法的句子。这就强烈地表明,句子的生成和理解有独立的模块——这样就能解释为什么丧失了后一种能力并不必然引起前一种能力的丧失。另外一些脑损伤的病人失去了长期记忆(遗忘症),但是短期记忆以及说话和理解能力丝毫没有受损。这似乎再次支持了模块性观点而反驳了把意识看成万能解题器的观点。

这种神经心理学上的证据尽管很有说服力,却没有一劳永逸地解决模块性的问题。一方面,这种证据比较稀少——显然不能只是为了了解认知能力受影响的状况而随意损坏人脑。另一方面,正如在科学中通常存在的,关于数据应该如何解释存在着严重的分歧。一些人认为,所观察到的脑损伤病人的认知障碍模式并不意味着意识是模块性的。他们声称,即使意识是万能解题器,即不是模块性的,脑损伤不同程度地影响不同的

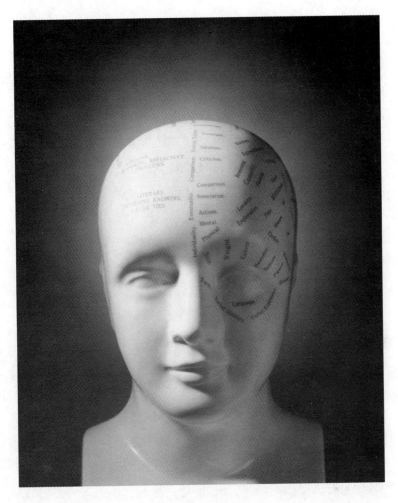

图 16 一种假设性的模块化意识示意图。

认知能力仍然是可能的。所以他们主张不能仅从缺陷研究来"轻率判断"意识的结构,这种研究最多只能提供有瑕疵的证据。

最近许多对模块性的关注要归功于杰里·福多尔,一位有影响力的美国哲学家和心理学家。福多尔于 1983 年出版了《意识的模块性》一书,该书既有对模块究竟为何物的非常清晰的论述,也有对哪些认知能力是模块化的、哪些不是的有趣假设。福多尔认为大脑模块有大量突出的特征,下面是其中最重要的三个特征:(i) 它们是**领域化的**;(ii) 它们的运行是**强制性的**,(iii)它们是**信息分隔的**。非模块化的认知系统不具有其中任何一个特征。福多尔接着主张,人类意识虽非全部但却部分是模块化的:有些认知任务我们用专门的模块来解决,有些任务我们用"通用智能"来解决。

说一个认知系统是领域化的,就是说它是专门化的:它负责一组有限的、精确划定的任务。乔姆斯基所假定的"语言习得机制"就是领域化系统的一个很好的例子。这种机制的唯一功能就是使儿童学会语言——它并不帮儿童学会下棋、数数或者做其他任何事。所以这种机制完全忽略非语言性的输入。说一个认知系统是强制性的,就是说我们不能选择是否让该系统运作。语言的感知是一个很好的例子。如果你听到一句用你所通晓的语言说出的句子,你就不得不把它听成是说出了一个句子。如果有人要你把该句听成"纯粹的噪音",不论如何努力,你都无法做到。福多尔指出,并非所有认知过程在这方面都是强制性的。**思维**明显就不是这样。如果有人让你回想生命中最恐惧的时刻,或者让你想象中了彩票后最想做的事,你明显能够照做。所以思维和语言感知在这方面非常不同。

信息分隔，即心理模块的第三个也是最重要的特征，又是怎样的呢？有个例子能最好地解释这一概念。观察一下图 17 中的两条线。

　　上面的那条线在多数人看来要比底下的那条长一点。但实际上这是一种视觉上的错觉，称做米勒－利耶尔错觉。实际上这两条线一样长。对于为什么上面那条线看起来更长，有着多种解释，这些解释并不是我们在此要关注的。这里的关键在于：**即使知道它是一种视觉幻觉**，这两条线看起来仍然不一样长。在福多尔看来，这一简单的事实对于理解意识结构有着重要的启发。它表明，关于两条线不一样长的信息已被存储在认知意识的一块区域中，这块区域是我们的感知机制所不能达到的。这就意味着我们的感知机制是信息分隔的——它们不能获得我们拥有的所有信息。如果视觉感知不是以这种方式被信息分隔，而是能够使用我们存储在意识中的所有信息，那么只要被告知这两条线实际上一样长，这种错觉就会消失。

　　信息分隔的另外一个可能的例子来自人类恐惧症的现象。拿恐蛇症，或者说对蛇的恐惧的例子来说，这种恐惧症在人类中非常普遍，在许多其他的灵长类动物中亦然。这容易理解，因为蛇对于灵长类动物来说非常危险，所以通过自然选择，就很

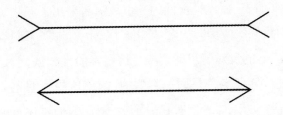

图 17　米勒－利耶尔错觉。两条横线在长度上是相等的，但是上面的那条看起来更长。

容易进化出对蛇的本能恐惧。但是不管对我们为什么这么怕蛇如何进行解释,关键之处仍在于下面这一点:即使你知道特定的一种蛇没有危险性,例如已经知道它的毒腺已被除去,你仍然很可能害怕这条蛇,而且不愿意触摸它。当然,这种恐惧症通常能通过训练来克服,但那是另外一回事。这里相关的要点是,该蛇不危险的信息不能进入你意识的这一部分,该部分在你看到蛇时会引起害怕的反应。这说明,每个人身上可能都有与生俱来的、信息分隔式的"恐蛇"模块。

你也许想知道为什么意识的模块性问题在根本上是一个哲学问题。意识是否是模块化的,这是否真的只是个经验事实的问题,尽管不容易回答?实际上这种说法不是很正确。模块性争论在一个方面是哲学性的,该方面关系到我们该怎么看待认知任务和认知模块。赞成模块性的人认为意识包含有执行不同认知任务的特定模块;反对模块性的人否定这一点。但是我们如何判定两个认知任务是同一类还是不同类呢?脸部识别是单一的认知任务还是由两个不同的认知任务构成的:识别男性的脸和识别女性的脸?做长除法和乘法是不同的认知任务,还是都是更一般的算术运算任务的一部分? 这类问题是概念上的,或者说是哲学上的,而不是直接经验上的,它们对于模块性争论可能非常重要。假设模块性的一位反对者提出了一些实验性证据,表明我们仅使用同一套认知能力来执行许多不同类型的认知任务。她的反对者可能会接受这些实验性数据,但是同时声称,相关认知任务都是**同**一类型的,因此这些数据完全与模块性相符合。所以尽管乍看起来不然,意识的模块化争论还是深陷在哲学争论中。

最热衷地赞成模块性的人相信意识完全由模块构成,但是这种观点并不被广泛接受。福多尔本人也认为，感知和语言很可能是模块化的，但思想和推理几乎肯定不是。为什么不是？假设你正参加陪审团，在决定是宣告有罪还是无罪裁决。你将怎样处理这一任务？你会考虑的一个重要的问题是，被告的陈述在逻辑上是否一致——是否没有矛盾？你可能问自己，现有的证据是否刚好与被告的罪行相符，或者是否很强地支持了罪行的成立。显然你在此所用的推理技巧——检查逻辑一致性和评估证据——是**通用的**技巧；它们不是专门设计出来用于陪审团的。你在许多领域都使用这些技巧。所以你在仔细考虑被告的罪行时所运用的认知能力不是领域化的。同样它们的运用也不是强制性的——你必须有意识地思考被告是否有罪，并且能够在任何你想要停止的时刻，例如在午休时间，停止这种思考。最重要的是，这里同样也没有信息分隔。你的任务是**全面考虑**，决定被告是否有罪，所以你也许必须运用所拥有的任何背景信息，只要你认为相关。例如，如果被告在审问之下紧张痉挛，并且你相信紧张的痉挛总是有罪的一种标志，你就可能会利用这一信念来作出裁决。所以这里没有信息的储存，它是你用来作出裁决的认知机制所不能通达的（尽管法官可能会提醒你忽视某些事情）。简言之，这里不存在决定一名被告是否有罪的模块。你是用"通用智能"解决这一认知问题的。

福多尔的命题，即意识尽管不是全部但部分是模块化的，这样看来便十分合理。但是确切说来有多少模块、这些模块具体负责什么，在当前的研究状况下还是无法回答。福多尔本人对认知心理学解释人类意识运作方式的的可能性非常悲观。他坚信，只有对模块化的系统才能进行科学地研究——非模块化系统因并非信息分隔而更难以做出模型。所以在福多尔看来，认知心理学家最好的研究策略是关注感知和语言,而不管思维和推理。但是福多尔思想的这个方面颇具争议。并非所有心理学家在意识的哪些部分是模块化的、哪些不是的问题上都同意他的观点,也并非所有心理学家都赞同,只有模块化的系统能够被科学地研究。

第七章
科学和科学批评者

　　由于显而易见的原因，许多人理所当然地认为科学是好的。毕竟，科学给我们带来了电力、安全的饮用水、盘尼西林、避孕方法、空中旅行，等等等等——所有的这些毫无疑问都已经使人类获益。但是除了这些对人类幸福的重要贡献之外，科学也受到了批评。有些人认为，社会以牺牲文化艺术为代价在科学上投入了过多的金钱；另一些人认为，科学赋予了我们在不拥有的情形下反而会过得更好的技术能力，诸如制造大规模杀伤性武器的能力（见图18）。某些女权主义者则认为科学是令人讨厌的，因为它内在地具有男子主义的偏见；那些具有宗教信仰的人经常感到科学威胁着他们的信仰；人类学家谴责西方科学的自负，理由是它漫不经心地认为自己凌驾于全世界各地本土文化的知识和信仰之上。这绝没有穷尽科学所遭受的所有批评，然而在本章中我们只关注三种具有特殊哲学意义的批评。

科学主义

　　当今时代，"科学"和"科学的"这两个词已经获得了空前的威望。如果有人指责你行为"不科学"，他们几乎肯定是在批评

117

你。科学的行为是明智、理性并且值得赞扬的；不科学的行为是愚蠢、非理性并且应该被鄙视的。很难去知晓为什么"科学的"这个标签获得了这些内涵，它有可能与科学在现代社会中所取得的崇高地位有关系。社会视科学家为专家，在重要的问题上请教他们，并且他们的绝大多数观点被毫无疑问地接受下来。当然，每一个人都意识到科学家有时也会犯错——例如，20世纪90年代英国政府的科学顾问宣称"疯牛病"不会威胁人类，恰恰悲剧性地被证明是错误的。但是这种偶然的失误并不会动摇公众对于科学的信念，也不会削弱科学家们所获得的尊重。至少在西方，科学家被看成是与过去的宗教领袖一样：凡夫俗子无法获得的专门知识的拥有者。

　　"科学主义"是一个带有贬义的标签，被一些哲学家用来描述他们眼中的科学崇拜——在许多知识领域发现的对于科学过于尊敬的态度。科学主义的反对者们认为，科学并不是知识探索的唯一有效形式，也不是通往知识的独一无二的优先通

图 18　如果没有科学的能力我们会过得更好：一次原子弹爆炸产生的有毒蘑菇云。

道。他们经常强调他们并不是反对科学**本身**,他们反对的是在现代社会中科学,尤其是自然科学所拥有的特权地位,以及科学方法必然能够适用于每一个学科的设想。所以他们的目的不是攻击科学而是摆正它的位置——表明科学仅是同等事物中的一种,并且把其他学科从可能凌驾于它们之上的科学专制中解放出来。

科学主义显然是一种相当模糊的说法,并且由于这一术语事实上已被滥用,几乎任何人都不承认会相信它。尽管如此,同科学崇拜颇类似的情形却是知识领域的一个真实情况。这并不必然是一件坏事——也许科学就应该受到崇拜。但是它确实是一个真实的现象。经常被谴责为科学崇拜的一个领域是当代英美哲学(科学哲学仅是其一个分支)。在传统的意义上,尽管与数学和科学有紧密的历史联系,哲学仍被认为是人文学科,并且具有很好的理由。因为哲学所探讨的问题包括知识、道德、理性、人类幸福等等的本质,它们中任何一个看起来都不能用科学的方法去解决。科学的任何一个分支都不会告诉我们应该如何生活、知识是什么或人类的幸福包含着什么;这些都是经典的哲学问题。

尽管明显不可能通过科学回答哲学的问题,相当多的当代哲学家却坚信科学是获得知识的唯一正当途径。他们认为,不能用科学方式解决的问题根本不是真正的问题。这种观点经常和最近去世的维拉德·范·奥曼·奎因联系在一起,他可能是美国20世纪最重要的哲学家。这种观点的根据在于一个名为"自然主义"的学说,该学说强调我们人类是自然界的主要部分,而不是如人们以前所认为的那样与其相分离。既然科学研究的是整个自然界,它是否必然有能力揭示出关于人类状况的全部真理,不会把任何东西留待哲学去揭示? 这种观点的拥护者有时补充说,科学毫无疑问不断进步,然而哲学却好像连续数个世纪讨论同一些问题。在这种观念下,并不存在明显属于哲学的知识,所有的知识都是科学知识。至于哲学独一无二的角色,就是"澄清科学概念"——清理好(工具)以便科学家能够继续进行研究。

意料之中,许多哲学家反对使他们的学科臣服于科学;这是反对科学主义的主要来源之一。他们认为,哲学探究揭示的真理是科学无法企及的。哲学问题不能通过科学的方式来解决,这无异于表明:科学不是获得真理的唯一途径。这种观点的支持者承认,在不提出与科学教给我们的东西相矛盾的观点的意义上,哲学应该努力同科学保持一致。并且他们承认,科学应该受到极其尊重的对待。他们所反对的是科学帝国主义——认为科学有能力回答所有的关于人类及其在自然中所处位置的重要问题。这种观点的倡导者通常也把自己看做自然主义者。他们通常不认为人类在某种程度上处于自然秩序之外,因而不是科学研究的对象。他们认为我们仅仅是另外一种生物种类,我们的身体最终都是由物理微粒构成的,正如宇宙中的一切。但是他们否认这意味着科学方法适合解释每一个人们感兴趣的问题。

围绕着自然科学与社会科学的关系,产生了一个类似的争论。正如哲学家有时抱怨他们学科中的"科学崇拜"那样,社会科学家有时也抱怨自己学科中的"自然科学崇拜"。毫无疑问自然科学——物理学、化学、生物学等——比社会科学——经济学、社会学、人类学等——发展得更为成熟。许多人对于这种状况迷惑不解。如果说是因为自然科学家比社会科学家更聪明,这几乎是不可能的。一个可能的答案是,自然科学的**方法**优于社会科学的方法。如果这种回答是正确的,社会科学为了赶上自然科学所需要做的就将是模仿自然科学的方法。在某种程度上,这种情况确实已经发生了。在社会科学中越来越多地使用数学,也许在某种程度上是这种观点的一个结果。当伽利略开始运用数学语言来描述运动时,物理学产生了一个巨大的飞跃;因此,人们很容易认为,只要能找到对社会科学的主题进行"数学化"的类似方法,社会科学领域就能实现类似的飞跃。

但是,一些社会科学强烈拒绝这种主张,即他们应该以这种方式仰望自然科学,正如某些哲学家强烈地拒绝仰望作为一个整体的科学。他们认为,自然科学的方法并不必然适用于研究社会现象。例如,在天文学上有用的同一种技术方法为什么对于研究社会具有同等的用处?持这种观点的那些人否认自然科学研究的日趋成熟可以归功于他们采用的独特探究方法,因此看不到任何理由应该把那些方法延伸到社会科学领域。他们经常指出,社会科学比自然科学年轻,社会现象的复杂本质使得成功的社会科学研究难以实现。

关于自然科学和社会科学地位等同的争议以及科学主义的争议都不容易解决。部分是因为,"科学方法"或者"自然科学方法"的确切内容远不够清晰——这一点被争论的双方经常忽略。如果我们想知道科学的方法是否可以应用于每一个学科,或者说它们是否有能力回答每一个重要的问题,我们显然需要知道那些方法确切**是**什么。但是正如我们在前面章节所看到的,这远远不如看上去那样简单。我们当然知道科学探究的一些主要特征:归纳、实验验证、观察、理论建构、最佳解释推论,等等。但是这一清单并没有为"科学方法"提供一个精确的定义。这样一个定义是否**能够**被提出,也并不清楚。科学发展日新月异,设想存在着一种被所有科学学科一直使用的固定不变的"科学方法",这远非理所当然。但是这种设想既包含在科学是获得知识的一条正确路径的观点中,**也**包含在相反的观点之中,即有些问题不能通过科学方法来解答。这就表明,至少在某种程度上,关于科学主义的争论可能在前提上就错了。

科学和宗教

科学和宗教之间的紧张由来已久,留下了大量纪录。也许最著名的案例要算伽利略和天主教的冲突。1633 年宗教裁判所强迫伽利略公开放弃哥白尼观点,并且判罚他在佛罗伦萨软禁中度过余生。天主教拒绝接受哥白尼的学说,当然是因为它与圣经相抵触。当代,在美国科学和宗教最显著的冲突是达尔文主义者和神创论者之间的激烈争论,这正是本节所要关注的。

达尔文进化论受到神学上的反对并不是什么新鲜的事。1859 年发表之时,《物种起源》就立即受到了来自英国教会人士的批判。原因很明显:达尔文的理论认为所有现存的物种,包括人类,都是从共同的祖先经过一个漫长的时间演变而来的。这种理论显然与圣经《创世记》矛盾,圣经上说上帝用六天的时间创造了所有的生物。所以抉择看起来是赤裸裸的:要么相信达尔文,要么相信圣经,但是不能二者都相信。尽管如此,许多忠于达尔文学派的人士已经找到方法来调和他们的基督教信仰和他们对于进化论的信念——其中包括许多杰出的生物学家。一个途径是,干脆不要对于冲突思考太多。另一个在理智上更为诚实的途径是,指出圣经不应该在字面上解释——它应该被看做是寓言性的或者象征性的。因为毕竟,达尔文的理论与上帝的存在以及基督教的其他许多教义具有相当的兼容性。达尔文主义排除的仅仅是圣经创世故事的表面事实。因此一种适当缩减的基督教教义就可与达尔文主义的相容。

但是,在美国,特别是南部诸州,许多福音教派的新教徒一直不愿改变宗教信仰来适应科学发现。他们坚持认为圣经的创世解释正确无误,达尔文的进化论是完全错误的。这种观点被称为"神创论",美国约有 40%的成年人接受,远远大于英国和欧洲大陆的比例。上帝创世说具有强大的政治力量,并且令科学家们非常沮丧的是,它对于美国中小学的生物学教学产生了重要影响。20 世纪 20 年代著名的"猴子审判"一案中,一位田纳西州的学校教师由于向学生教授进化论被判违反了州法。(这一法律最终在 1967 年被联邦最高法院所推翻。)部分由于猴子审判,进化论的课程完全从美国高中的生物学科目中被取消了数十年。美国的几代成年人都是在对达尔文一无所知的情况下长大的。

20 世纪 60 年代,情况发生了改变,神创论者与达尔文主义者之间又起争执,并且引发了被称为"神创科学"的运动。神创论者想要中学学生完全按照《创世记》中所讲的那样学习圣经的神创故事。但是美国宪法禁止在公立学校里教授宗教知识。神创科学的概念正是为了绕过这一点。它的发明者认为,圣经中创世的解释为地球上的生命提供了一个比达尔文进化论更科学的解释。所以教授圣经的创世说并不违反宪法的禁令,因为它是作为科学而不是宗教进行讲授的!在整个南方腹地,产生了在生物学课堂上讲授神创科学的要求,这些要求常能受到关注。1981 年阿肯色州通过了一个法律,要求生物学老师给予神创科学和进化论"等同的授课时间",其他的州也相继效仿。尽管阿肯色州的这一法律在 1982 年被一位联邦法官判定为违宪,但是一直到今天给予"等同授课时间"的呼声仍然可以听到。它经常被称为一个公平的妥协——面对两套矛盾的信仰,还能有什么比给予每一个信仰等同的时间更公平的呢?民意调查显示,绝大多数的美国成年人都赞同:他们希望神创科学在公立学校中与进化论同样得到讲授。

但是,实际上所有的职业生物学家都把神创科学看做是一个借口——一种在科学的伪装下推进宗教信仰的不诚实且不正确的企图,最终会带来极其有害的教育后果。为了与这种反对意见抗衡,神创论的科学家们已经付出巨大努力来削弱达尔文主义。他们认为达尔文主义的证据非常苍白,所以达尔文主义不是已成立的事实,而仅仅是一个理论。另外,他们还关注到了达尔文学派内部之间的各种分歧,对于个别生物学家的一些不谨慎言论进行挑剔,以显示对于进化论的不同意见在科学上是可敬的。他们得出的结论是,既然达尔文主义"仅是一个理论",学生就应该接触到其他的不同理论——诸如上帝六天创世的神创论。

在某种程度上,神创论者指出达尔文主义"仅仅是一个理论"而非被证明的事实是完全正确的。如我们在第二章所了解的,在严格意义上,**证明**一个科学理论正确是永远不可能的,因为从数据到理论的推论总是非演绎性的。但这只是一个普遍性的观点——它和进化论**本身**毫无关系。出于同样的原因,我们可以认为地球围绕太阳旋转、水是由 H_2O 组成的或者没有支撑的物体将会掉落等也"仅仅是一个理论",因此这些理论的对立观点学生都应该能接触到。但是神创论科学家并不这么认

科学哲学

124

为。他们怀疑的不是整体意义上的科学,而是特殊意义上的进化论。所以他们的立场要站得住脚,就不能简单地诉诸数据资料并不保证达尔文理论的正确性这种观点。因为每一个科学理论都是如此,事实上大多数常识信念也是。

客观地说,神创论科学家们的确提供了关于进化论的独特观点。他们最偏爱的一个观点是,化石记录非常不完全,特别是关于推想的**智人**祖先的。这一指责中有一些正确的成分。进化论者们很长时间都困惑于化石记录上的代际差距。一个一直以来的困惑是:为什么只有那么少的"过渡型化石"——两个物种之间中介的生物化石。如果按照达尔文理论的断言,较晚的物种由较以前的物种进化而来,我们就一定可以期待过渡化石非常普遍吗?神创论者利用这种困惑来表明,达尔文的理论恰恰是错误的。但是神创论者的论证并没有说服力,尽管在理解化石记录的问题上确实存在困难。因为化石并不是唯一的或者不是主要的进化论证据来源,神创论者如果读过《物种起源》,就会了解这一点。比较解剖学科是另外一个重要的证据来源,如胚胎学、生物地理学和遗传学。例如,来看一个事实:人类和黑猩猩有 98% 的 DNA 是相同的。如果进化论是正确的,这种以及成千上万种类似的事实就会变得非常有意义,并且由此成为支持进化论的极好证据。当然,神创论科学家也可以解释这样的事实。他们可以声称:上帝出于**他**自己的原因,决定使人类和黑猩猩在基因上相似。然而,给出这种"解释"的可能性的确正好指出了一个事实:达尔文的理论在逻辑上并非必然能由数据推出的。正如我们已经了解的,每一种科学理论都面临着这样的问题。神创论者仅仅强调了普遍的方法论观点,即数据总是能够通过多种方式得到解释。这一观点是正确的,但是并没有表明什么是与达尔文主义特别相关的。

尽管神创论科学家的观点普遍没有根据，但是神创论者／达尔文主义者之间的争议确实提出了关于科学教育的重要问题。在世俗的教育体系中应该如何处理科学和信仰之间的冲突？应该由谁来决定中学科学课堂的内容？纳税人对于靠税收支持的学校所教授的内容有发言权吗？不想让孩子接授进化论或者其他科学内容的父母应该遭到国家的否决吗？诸如此类的公共政策事务通常很少得到讨论，但是达尔文主义者和神创论者之间的冲突却把它们推到了聚光灯之下。

科学是价值无涉的吗？

几乎每个人都会同意，科学知识有时被用在了不道德的目的上——例如用于制造核武器、生物武器和化学武器。但是这些例子并不表明科学知识本身在伦理道德上应该遭到反对。不道德的是知识的**使用**。事实上，很多哲学家会说关于科学或科学知识**本身**是道德还是非道德的讨论是没有意义的。科学所涉及的是事实，而事实本身是没有伦理意义的。有对错之分以及道德和不道德差别的是我们对这些事实的所作所为。以这种观点来看，科学本质上是**价值无涉**的活动——科学的任务仅在于提供关于世界的信息。社会愿意用这种信息来做什么是另一回事。

并非所有哲学家都承认科学在价值问题上是中立的，也不承认这种主张所依据的事实和价值二分法。一些哲学家声称价值中立的理想是不可能达到的——科学探究总是渗透了价值判断。（这与第四章中所讨论的所有观察都负荷了理论的观点相类似。实际上，这两个观点通常是关联在一起的。）反驳价值无涉科学之可能性的一个观点来自这样一个明显的事实,即科学家必须选择研究什么——不能同时研究所有事物。所以必须判断可能研究的不同对象间的相对重要性,在弱的意义上说,这就是价值判断。另外一个反驳观点源于诸君现在应该已经熟知的一个事实,即任何一组数据原则上都可以通过一种以上的方式来解释。这样,科学家的理论选择就绝不会单由数据来决定。一些哲学家用这一点来表明,价值是不可避免地包含在理论选择中的,因此科学不可能是价值无涉的。第三个反驳观点是,科学知识不能如价值中立所要求的那样与它的预期应用脱离开来。以此观点来看,把科学家描绘成不考虑研究的实际应用而无私地为了研究而研究是很天真的。当今的许多科学研究都是由私营企业资助,这些私营企业明显拥有既得的商业利益——这一事实为该观点提供了一些证据。

这些观点虽然有趣,却都有些抽象——它们试图表明科学在原则上不可能价值无涉,却没有指出价值介入科学的实际案例。但是,对价值负荷的个案指责也已经出现。其中一个例子与被称为人类社会生物学的学科相关,该学科在 20 世纪 70 年代和 80 年代产生了大量的争论。人类社会生物学试图把达尔文理论的原理应用到人类行为上。这种方案乍一看来非常合理。因为人类只是动物的一个物种,并且生物学家承认达尔文理论能够解释大量的动物行为。例如,对于为什么老鼠在看到猫时通常会逃走,达尔文主义就有很清楚的解释。过去,不逃走的老鼠比逃走的老鼠往往留下更少的后代,因为它们被猫吃了;假设这种行为是基于基因的,并因而从上代传到下代,许多代以后这种逃跑行为就会遍及该种群。这就解释了为什么现在的老鼠要避开猫。这种解释就是所谓的"达尔文主义的"或"适应主

义的"解释

人类社会生物学家(以下简称"社会生物学家")坚信,人类的许多行为特征都可以给出适应主义的解释。他们最喜欢的一个例子是乱伦回避。乱伦——或者同一家庭成员间的性关系——在几乎每个人类社会中都被视为禁忌,并且在多数社会中都要受到法律和道德的制裁。这一事实非常令人惊讶,因为整个人类社会中有关性的风俗习惯极为不同。为什么要禁止乱伦? 社会生物学家作了以下解释:通过乱伦关系生出的孩子通常有严重的基因缺陷。所以在过去,那些乱伦的人比不乱伦的人往往留下更少的能够存活下来的后代。假设乱伦回避行为是基于基因的,并因而从上代传到下代,那么许多代以后这种回避行为就会遍及人类。这就解释了为什么在现今的人类社会中极少能发现乱伦现象。

很容易理解,许多人对这种解释感到不安。因为,社会生物学家实际上在说我们回避乱伦是在基因上就被先天决定了的。这有悖于我们的常识观点,即回避乱伦是因为我们被告知这是错误的, 也就是说我们的行为具有文化上而非生物学上的解释。回避乱伦实际上属于少有争议的例子之一。社会生物学家作出适应主义解释的其他行为包括强奸、侵犯行为、仇外和男性乱交。在每个例子中,他们都采用了相同的论证:从事这种行为的个体繁殖能力超过不从事这种行为的个体,该行为是基于基因的,所以它就从上代传到了下代。当然,并非所有人都是有侵犯性的、仇外的或者是进行过强奸的。但是这并不表明社会生物学家是错的。因为他们的论证仅要求这些行为有基因成分,即存在着某种或某些这样的基因,它们能提高基因携带者从事这些行为的概率。这比说行为完全由基因决定(这种说法几乎肯定是错的）措辞要弱得多。换言之，社会生物学的描述是要解释，为什么人类之中有侵犯性、仇外性和强奸别人的**倾向**——尽管这些倾向很少表现出来。所以侵犯、恐外和强奸（谢天谢地）非常少见这一事实，本身并不证明社会生物学家是错的。

社会生物学遭到了来自众多领域学者的强烈批判。其中一些是严格科学意义上的。批判指出社会生物学的假设极难检验，因而应被视做有趣的猜测，而不是确定的真理。然而其他批判反对得更为根本，认为整个社会生物学的研究纲领在思想意识上都令人怀疑。他们把它看做是试图为通常由男性作出的反社会行为辩护或开脱。例如，通过论证强奸有基因成分，社会生物学家意指强奸是"自然的"，因而强奸者对自己的行为并不真正负有责任——他们只是服从了基因冲动。社会生物学家似乎在说："如果强奸行为责任在于基因，我们又怎能谴责强奸者呢？"仇外和男性乱交的社会生物学解释被看做是同样有害的。他们似乎在暗示，种族歧视和婚姻不忠之类被多数人视为不良的现象，是自然的和不可避免的——是基因遗传的产物。简言之，批评者指责社会生物学是负荷了价值的科学，并且它负荷的价值非常成问题。或许不足为奇的是，这些批评者中有许多女权主义者和社会科学家。

对于这种指责的一个可能的回应是坚持事实与价值的区分。以强奸为例，按推测来看，或者有一种使男人倾向于强奸的、通过自然选择而扩散开来的基因，或者没有这样一种基因。这是一个纯科学事实的问题，尽管是个不易解答的问题。然而事实是一回事，价值又是另外一回事。即使存在这样一种基因，它也并不能使强奸可被原谅或可被接受。它同样不会使强奸者为他们的行为少负责任，因为没有人认为这种基因能真正地**迫使**男性去强奸。至多，这种基因会预先使男人倾向于强奸，但是天生的倾向性能够通过文化的教化来克服，并且每个人都被教导强奸是错的。这同样可应用到仇外、侵犯行为和乱交上。即使这些行为的社会生物学解释是正确的，它对于我们管理社会或者任何其他的政治或伦理事务也没有意义。伦理学不能从科学推导出来。所以关于社会生物学，没有什么在思想意识上要受质疑。与所有科学一样，它仅仅试图告诉我们关于世界的事实。有时事实令人不安，但我们必须学会接受。

如果这种回应是正确的，它就意味着我们应该严格区分对于社会生物学的"科学的"反驳与"思想意识上的"反驳。尽管这听起来合理，但是有一点却没有提及：社会生物学的倡导者在政治上倾向于右翼，而它的批评者往往是政治上的左翼分子。对于这一归纳，有很多例外，尤其是前者，但是几乎没有人会否认这一总的倾向。如果社会生物学只不过是对事实的一种没有偏见的探究，这种倾向又作何解释？为什么在政治观点和对社会生物学的看法之间应有任何的关联？这是一个非常棘手的问题。因为尽管一些社会生物学家或许具有掩藏的政治关怀，并且其批判者具有属于自己的相对立的政治主张，但是这种关联甚至延伸到了那些用明显的科学术语争论问题的人那里。这表明，尽管没有证明，"思想意识上的"与"科学的"问题也许不那么容易彻底分开。所以，社会生物学是否价值无涉的问题要比原来所想象的复杂。

总而言之，像科学这样在现代社会中充当着关键角色并且耗费了如此多公共财政的事业，必然会受到来自多种渠道的批判。这也是一种好事，因为不加批判地接受科学家所说和所做的每件事将是既危险又武断的。可以有把握地预言, 21 世纪的科学,通过它的技术应用将会比过去在更大程度上影响日常生活。所以"科学是个好东西吗?"这个问题将仍会变得更为紧迫。哲学上的反思也许不会就此问题得出一个最终的、明确的答案,但是它有助于分离出关键的要点,并且促进对它们进行合理的、平和的讨论。

索引
（条目后的数字为原文页码）

W

Samir Okasha

PHILOSOPHY OF SCIENCE

A Very Short Introduction

Contents

Acknowledgements

I would like to thank Bill Newton-Smith, Peter Lipton, Elizabeth Okasha, Liz Richardson and Shelley Cox for reading and commenting on earlier versions of this material.

Samir Okasha

List of illustrations

Chapter 1
What is science?

What is science? This question may seem easy to answer: everybody knows that subjects such as physics, chemistry, and biology constitute science, while subjects such as art, music, and theology do not. But when as philosophers we ask what science is, that is not the sort of answer we want. We are not asking for a mere list of the activities that are usually called 'science'. Rather, we are asking what common feature all the things on that list share, i.e. what it is that *makes* something a science. Understood this way, our question is not so trivial.

But you may still think the question is relatively straightforward. Surely science is just the attempt to understand, explain, and predict the world we live in? This is certainly a reasonable answer. But is it the whole story? After all, the various religions also attempt to understand and explain the world, but religion is not usually regarded as a branch of science. Similarly, astrology and fortune-telling are attempts to predict the future, but most people would not describe these activities as science. Or consider history. Historians try to understand and explain what happened in the past, but history is usually classified as an arts subject not a science subject. As with many philosophical questions, the question 'what is science?' turns out to be trickier than it looks at first sight.

Many people believe that the distinguishing features of science lie in

1

the particular methods scientists use to investigate the world. This suggestion is quite plausible. For many sciences do employ distinctive methods of enquiry that are not found in non-scientific disciplines. An obvious example is the use of experiments, which historically marks a turning-point in the development of modern science. Not all the sciences are experimental though – astronomers obviously cannot do experiments on the heavens, but have to content themselves with careful observation instead. The same is true of many social sciences. Another important feature of science is the construction of theories. Scientists do not simply record the results of experiment and observation in a log book – they usually want to explain those results in terms of a general theory. This is not always easy to do, but there have been some striking successes. One of the key problems in philosophy of science is to understand how techniques such as experimentation, observation, and theory-construction have enabled scientists to unravel so many of nature's secrets.

The origins of modern science

In today's schools and universities, science is taught in a largely ahistorical way. Textbooks present the key ideas of a scientific discipline in as convenient a form as possible, with little mention of the lengthy and often tortuous historical process that led to their discovery. As a pedagogical strategy, this makes good sense. But some appreciation of the history of scientific ideas is helpful for understanding the issues that interest philosophers of science. Indeed as we shall see in Chapter 5, it has been argued that close attention to the history of science is indispensable for doing good philosophy of science.

The origins of modern science lie in a period of rapid scientific development that occurred in Europe between the years 1500 and 1750, which we now refer to as the scientific revolution. Of course scientific investigations were pursued in ancient and medieval times too – the scientific revolution did not come from nowhere. In these earlier periods the dominant world-view was Aristotelianism, named after the ancient Greek philosopher Aristotle, who put forward detailed theories in physics, biology, astronomy, and cosmology. But Aristotle's ideas would seem very strange to a modern scientist, as would his methods of enquiry. To pick just one example, he believed that all earthly bodies are composed of just four elements: earth, fire, air, and water. This view is obviously at odds with what modern chemistry tells us.

The first crucial step in the development of the modern scientific world-view was the Copernican revolution. In 1542 the Polish astronomer Nicolas Copernicus (1473–1543) published a book attacking the geocentric model of the universe, which placed the stationary earth at the centre of the universe with the planets and the sun in orbit around it. Geocentric astronomy, also known as Ptolemaic astronomy after the ancient Greek astronomer Ptolemy, lay at the heart of the Aristotelian world-view, and had gone largely unchallenged for 1,800 years. But Copernicus suggested an alternative: the *sun* was the fixed centre of the universe, and the planets, including the earth, were in orbit around the sun (Figure 1). On this heliocentric model the earth is regarded as just another planet, and so loses the unique status that tradition had accorded it. Copernicus' theory initially met with much resistance, not least from the Catholic Church who regarded it as contravening the Scriptures and in 1616 banned books advocating the earth's motion. But within 100 years Copernicanism had become established scientific orthodoxy.

Copernicus' innovation did not merely lead to a better astronomy. Indirectly, it led to the development of modern physics, through the work of Johannes Kepler (1571–1630) and Galileo Galilei (1564–1642). Kepler discovered that the planets do not move in circular orbits around the sun, as Copernicus thought, but rather in ellipses. This was his crucial 'first law' of planetary motion; his second and third laws specify the speeds at which the planets orbit the sun.

3

1. Copernicus' heliocentric model of the universe, showing the planets, including the earth, orbiting the sun.

Taken together, Kepler's laws provided a far superior planetary theory than had ever been advanced before, solving problems that had confounded astronomers for centuries. Galileo was a life-long supporter of Copernicanism, and one of the early pioneers of the telescope. When he pointed his telescope at the heavens, he made a wealth of amazing discoveries, including mountains on the moon, a vast array of stars, sun-spots, and Jupiter's moons. All of these conflicted thoroughly with Aristotelian cosmology, and played a pivotal role in converting the scientific community to Copernicanism.

Galileo's most enduring contribution, however, lay not in astronomy but in mechanics, where he refuted the Aristotelian theory that heavier bodies fall faster than lighter ones. In place of this theory, Galileo made the counter-intuitive suggestion that all

4

freely falling bodies will fall towards the earth at the same rate, irrespective of their weight (Figure 2). (Of course in practice, if you drop a feather and a cannon-ball from the same height the cannon-ball will land first, but Galileo argued that this is simply due to air resistance – in a vacuum, they would land together.) Furthermore, he argued that freely falling bodies accelerate uniformly, i.e. gain equal increments of speed in equal times; this is known as Galileo's law of free-fall. Galileo provided persuasive though not totally conclusive evidence for this law, which formed the centrepiece of his theory of mechanics.

Galileo is generally regarded as the first truly modern physicist. He was the first to show that the language of mathematics could be used to describe the behaviour of actual objects in the material world, such as falling bodies, projectiles, etc. To us this seems obvious – today's scientific theories are routinely formulated in mathematical language, not only in the physical sciences but also in biology and economics. But in Galileo's day it was not obvious: mathematics was widely regarded as dealing with purely abstract entities, and hence inapplicable to physical reality. Another innovative aspect of Galileo's work was his emphasis on the importance of testing hypotheses experimentally. To the modern scientist, this may again seem obvious. But at the time that Galileo was working, experimentation was not generally regarded as a reliable means of gaining knowledge. Galileo's emphasis on experimental testing marks the beginning of an empirical approach to studying nature that continues to this day.

The period following Galileo's death saw the scientific revolution rapidly gain in momentum. The French philosopher, mathematician, and scientist René Descartes (1596–1650) developed a radical new 'mechanical philosophy', according to which the physical world consists simply of inert particles of matter interacting and colliding with one another. The laws governing the motion of these particles or 'corpuscles' held the key to understanding the structure of the Copernican universe, Descartes

5

'*They were seen to fall evenly.*'

2. Sketch of Galileo's mythical experiment on the velocity of objects dropped from the Leaning Tower of Pisa.

believed. The mechanical philosophy promised to explain all observable phenomena in terms of the motion of these inert, insensible corpuscles, and quickly became the dominant scientific vision of the second half of the 17th century; to some extent it is still with us today. Versions of the mechanical philosophy were espoused by figures such as Huygens, Gassendi, Hooke, Boyle, and others; its widespread acceptance marked the final downfall of the Aristotelian world-view.

The scientific revolution culminated in the work of Isaac Newton (1643–1727), whose achievements stand unparalleled in the history of science. Newton's masterpiece was his *Mathematical Principles of Natural Philosophy*, published in 1687. Newton agreed with the mechanical philosophers that the universe consists simply of particles in motion, but sought to improve on Descartes' laws of motion and rules of collision. The result was a dynamical and mechanical theory of great power, based around Newton's three laws of motion and his famous principle of universal gravitation. According to this principle, every body in the universe exerts a gravitational attraction on every other body; the strength of the attraction between two bodies depends on the product of their masses, and on the distance between them squared. The laws of motion then specify how this gravitational force affects the bodies' motions. Newton elaborated his theory with great mathematical precision and rigour, inventing the mathematical technique we now call 'calculus'. Strikingly, Newton was able to show that Kepler's laws of planetary motion and Galileo's law of free-fall (both with certain minor modifications) were logical consequences of his laws of motion and gravitation. In other words, the very same laws would explain the motions of bodies in both terrestrial and celestial domains, and were formulated by Newton in a precise quantitative form.

Newtonian physics provided the framework for science for the next 200 years or so, quickly replacing Cartesian physics. Scientific confidence grew rapidly in this period, due largely to the success of

Newton's theory, which was widely believed to have revealed the true workings of nature, and to be capable of explaining everything, in principle at least. Detailed attempts were made to extend the Newtonian mode of explanation to more and more phenomena. The 18th and 19th centuries both saw notable scientific advances, particularly in the study of chemistry, optics, energy, thermodynamics, and electromagnetism. But for the most part, these developments were regarded as falling within a broadly Newtonian conception of the universe. Scientists accepted Newton's conception as essentially correct; all that remained to be done was to fill in the details.

Confidence in the Newtonian picture was shattered in the early years of the 20th century, thanks to two revolutionary new developments in physics: relativity theory and quantum mechanics. Relativity theory, discovered by Einstein, showed that Newtonian mechanics does not give the right results when applied to very massive objects, or objects moving at very high velocities. Quantum mechanics, conversely, shows that the Newtonian theory does not work when applied on a very small scale, to subatomic particles. Both relativity theory and quantum mechanics, especially the latter, are very strange and radical theories, making claims about the nature of reality that many people find hard to accept or even understand. Their emergence caused considerable conceptual upheaval in physics, which continues to this day.

So far our brief account of the history of science has focused mainly on physics. This is no accident, as physics is both historically very important and in a sense the most fundamental of all scientific disciplines. For the objects that other sciences study are themselves made up of physical entities. Consider botany, for example. Botanists study plants, which are ultimately composed of molecules and atoms, which are physical particles. So botany is obviously less fundamental than physics – though that is not to say it is any less important. This is a point we shall return to in Chapter 3. But even a brief description of modern science's origins would be incomplete if it omitted all mention of the non-physical sciences.

In biology, the event that stands out is Charles Darwin's discovery of the theory of evolution by natural selection, published in *The Origin of Species* in 1859. Until then it was widely believed that the different species had been separately created by God, as the Book of Genesis teaches. But Darwin argued that contemporary species have actually evolved from ancestral ones, through a process known as natural selection. Natural selection occurs when some organisms leave more offspring than others, depending on their physical characteristics; if these characteristics are then inherited by their offspring, over time the population will become better and better adapted to the environment. Simple though this process is, over a large number of generations it can cause one species to evolve into a wholly new one, Darwin argued. So persuasive was the evidence Darwin adduced for his theory that by the start of the 20th century it was accepted as scientific orthodoxy, despite considerable theological opposition (Figure 3). Subsequent work has provided striking confirmation of Darwin's theory, which forms the centrepiece of the modern biological world-view.

The 20th century witnessed another revolution in biology that is not yet complete: the emergence of molecular biology, in particular molecular genetics. In 1953 Watson and Crick discovered the structure of DNA, the hereditary material that makes up the genes in the cells of living creatures (Figure 4). Watson and Crick's discovery explained how genetic information can be copied from one cell to another, and thus passed down from parent to offspring, thereby explaining why offspring tend to resemble their parents. Their discovery opened up an exciting new area of biological research. In the 50 years since Watson and Crick's work, molecular biology has grown fast, transforming our understanding of heredity and of how genes build organisms. The recent attempt to provide a molecular-level description of the complete set of genes in a human

MR. BERGH TO THE RESCUE.

THE DEFRAUDED GORILLA. "That *Man* wants to claim my Pedigree. He says he is one of my Descendants."

Mr. BERGH. "Now, Mr. DARWIN, how could you insult him so?"

3. Darwin's suggestion that humans and apes have descended from common ancestors caused consternation in Victorian England.

being, known as the Human Genome Project, is an indication of how far molecular biology has come. The 21st century will see further exciting developments in this field.

More resources have been devoted to scientific research in the last hundred years than ever before. One result has been an explosion of new scientific disciplines, such as computer science, artificial intelligence, linguistics, and neuroscience. Possibly the most significant event of the last 30 years is the rise of cognitive science,

4. James Watson and Francis Crick with the famous 'double helix' – their molecular model of the structure of DNA, discovered in 1953.

which studies various aspects of human cognition such as perception, memory, learning, and reasoning, and has transformed traditional psychology. Much of the impetus for cognitive science comes from the idea that the human mind is in some respects similar to a computer, and thus that human mental processes can be understood by comparing them to the operations computers carry out. Cognitive science is still in its infancy, but promises to reveal much about the workings of the mind. The social sciences, especially economics and sociology, have also flourished in the 20th century, though many people believe they still lag behind the natural sciences in terms of sophistication and rigour. This is an issue we shall return to in Chapter 7.

What is philosophy of science?

The principal task of philosophy of science is to analyse the methods of enquiry used in the various sciences. You may wonder why this task should fall to philosophers, rather than to the scientists themselves. This is a good question. Part of the answer is that looking at science from a philosophical perspective allows us to probe deeper – to uncover assumptions that are implicit in scientific practice, but which scientists do not explicitly discuss. To illustrate, consider scientific experimentation. Suppose a scientist does an experiment and gets a particular result. He repeats the experiment a few times and keeps getting the same result. After that he will probably stop, confident that were he to keep repeating the experiment, under exactly the same conditions, he would continue to get the same result. This assumption may seem obvious, but as philosophers we want to question it. *Why* assume that future repetitions of the experiment will yield the same result? How do we know this is true? The scientist is unlikely to spend too much time puzzling over these somewhat curious questions: he probably has better things to do. They are quintessentially philosophical questions, to which we return in the next chapter.

So part of the job of philosophy of science is to question assumptions that scientists take for granted. But it would be wrong to imply that scientists never discuss philosophical issues themselves. Indeed, historically, many scientists have played an important role in the development of philosophy of science. Descartes, Newton, and Einstein are prominent examples. Each was deeply interested in philosophical questions about how science should proceed, what methods of enquiry it should use, how much confidence we should place in those methods, whether there are limits to scientific knowledge, and so on. As we shall see, these questions still lie at the heart of contemporary philosophy of science. So the issues that interest philosophers of science are not 'merely philosophical'; on the contrary, they have engaged the attention of some of the greatest scientists of all. That having been said, it must be admitted that many scientists today take little interest in philosophy of science, and know little about it. While this is unfortunate, it is not an indication that philosophical issues are no longer relevant. Rather, it is a consequence of the increasingly specialized nature of science, and of the polarization between the sciences and the humanities that characterizes the modern education system.

You may still be wondering exactly what philosophy of science is all about. For to say that it 'studies the methods of science', as we did above, is not really to say very much. Rather than try to provide a more informative definition, we will proceed straight to consider a typical problem in the philosophy of science.

Science and pseudo-science

Recall the question with which we began: what is science? Karl Popper, an influential 20th-century philosopher of science, thought that the fundamental feature of a scientific theory is that it should be falsifiable. To call a theory falsifiable is not to say that it is false. Rather, it means that the theory makes some definite predictions that are capable of being tested against experience. If these predictions turn out to be wrong, then the theory has been falsified, or disproved. So a falsifiable theory is one that we might discover to be false – it is not compatible with every possible course of experience. Popper thought that some supposedly scientific theories did not satisfy this condition and thus did not deserve to be called science at all; rather they were merely pseudo-science.

Freud's psychoanalytic theory was one of Popper's favourite examples of pseudo-science. According to Popper, Freud's theory could be reconciled with any empirical findings whatsoever. Whatever a patient's behaviour, Freudians could find an explanation of it in terms of their theory – they would never admit that their theory was wrong. Popper illustrated his point with the following example. Imagine a man who pushes a child into a river with the intention of murdering him, and another man who sacrifices his life in order to save the child. Freudians can explain both men's behaviour with equal ease: the first was repressed, and the second had achieved sublimation. Popper argued that through the use of such concepts as repression, sublimation, and unconscious desires, Freud's theory could be rendered compatible with any clinical data whatever; it was thus unfalsifiable.

The same was true of Marx's theory of history, Popper maintained. Marx claimed that in industrialized societies around the world, capitalism would give way to socialism and ultimately to communism. But when this didn't happen, instead of admitting that Marx's theory was wrong, Marxists would invent an *ad hoc* explanation for why what happened was actually perfectly consistent with their theory. For example, they might say that the inevitable progress to communism had been temporarily slowed by the rise of the welfare state, which 'softened' the proletariat and weakened their revolutionary zeal. In this sort of way, Marx's theory could be made compatible with any possible course of events, just like Freud's. Therefore neither theory qualifies as genuinely scientific, according to Popper's criterion.

Popper contrasted Freud's and Marx's theories with Einstein's theory of gravitation, also known as general relativity. Unlike Freud's and Marx's theories, Einstein's theory made a very definite prediction: that light rays from distant stars would be deflected by the gravitational field of the sun. Normally this effect would be impossible to observe – except during a solar eclipse. In 1919 the English astrophysicist Sir Arthur Eddington organized two expeditions to observe the solar eclipse of that year, one to Brazil and one to the island of Principe off the Atlantic coast of Africa, with the aim of testing Einstein's prediction. The expeditions found that starlight was indeed deflected by the sun, by almost exactly the amount Einstein had predicted. Popper was very impressed by this. Einstein's theory had made a definite, precise prediction, which was confirmed by observations. Had it turned out that starlight was not deflected by the sun, this would have showed that Einstein was wrong. So Einstein's theory satisfies the criterion of falsifiability.

Popper's attempt to demarcate science from pseudo-science is intuitively quite plausible. There is certainly something fishy about a theory that can be made to fit any empirical data whatsoever. But some philosophers regard Popper's criterion as overly simplistic. Popper criticized Freudians and Marxists for explaining away any data that appeared to conflict with their theories, rather than accepting that the theories had been refuted. This certainly looks like a suspicious procedure. However, there is some evidence that this very procedure is routinely used by 'respectable' scientists – whom Popper would not want to accuse of engaging in pseudo-science – and has led to important scientific discoveries.

Another astronomical example can illustrate this. Newton's gravitational theory, which we encountered earlier, made predictions about the paths the planets should follow as they orbit the sun. For the most part, these predictions were borne out by observation. However, the observed orbit of Uranus consistently differed from what Newton's theory predicted. This puzzle was solved in 1846 by two scientists, Adams in England and Leverrier in France, working independently. They suggested that there was another planet, as yet undiscovered, exerting an additional gravitational force on Uranus. Adams and Leverrier were able to calculate the mass and position that this planet would have to have, if its gravitational pull was indeed responsible for Uranus' strange behaviour. Shortly afterwards the planet Neptune was discovered, almost exactly where Adams and Leverrier had predicted.

Now clearly we should not criticize Adams' and Leverrier's behaviour as 'unscientific' – after all, it led to the discovery of a new planet. But they did precisely what Popper criticized the Marxists for doing. They began with a theory – Newton's theory of gravity – which made an incorrect prediction about Uranus' orbit. Rather than concluding that Newton's theory must be wrong, they stuck by

15

the theory and attempted to explain away the conflicting observations by postulating a new planet. Similarly, when capitalism showed no signs of giving way to communism, Marxists did not conclude that Marx's theory must be wrong, but stuck by the theory and tried to explain away the conflicting observations in other ways. So surely it is unfair to accuse Marxists of engaging in pseudo-science if we allow that what Adams and Leverrier did counted as good, indeed exemplary, science?

This suggests that Popper's attempt to demarcate science from pseudo-science cannot be quite right, despite its initial plausibility. For the Adams/Leverrier example is by no means atypical. In general, scientists do not just abandon their theories whenever they conflict with the observational data. Usually they look for ways of eliminating the conflict without having to give up their theory; this is a point we shall return to in Chapter 5. And it is worth remembering that virtually every theory in science conflicts with some observations – finding a theory that fits all the data perfectly is extremely difficult. Obviously if a theory persistently conflicts with more and more data, and no plausible ways of explaining away the conflict are found, it will eventually have to be rejected. But little progress would be made if scientists simply abandoned their theories at the first sign of trouble.

The failure of Popper's demarcation criterion throws up an important question. Is it actually possible to find some common feature shared by all the things we call 'science', and not shared by anything else? Popper assumed that the answer to this question was yes. He felt that Freud's and Marx's theories were clearly unscientific, so there must be some feature that they lack and that genuine scientific theories possess. But whether or not we accept Popper's negative assessment of Freud and Marx, his assumption that science has an 'essential nature' is questionable. After all, science is a heterogeneous activity, encompassing a wide range of different disciplines and theories. It may be that they share some fixed set of features that define what it is to be a science, but it may not. The philosopher Ludwig Wittgenstein argued that there is no fixed set of features that define what it is to be a 'game'. Rather, there is a loose cluster of features most of which are possessed by most games. But any particular game may lack any of the features in the cluster and still be a game. The same may be true of science. If so, a simple criterion for demarcating science from pseudo-science is unlikely to be found.

Chapter 2
Scientific reasoning

Scientists often tell us things about the world that we would not otherwise have believed. For example, biologists tell us that we are closely related to chimpanzees, geologists tell us that Africa and South America used to be joined together, and cosmologists tell us that the universe is expanding. But how did scientists reach these unlikely-sounding conclusions? After all, no one has ever seen one species evolve from another, or a single continent split into two, or the universe getting bigger. The answer, of course, is that scientists arrived at these beliefs by a process of reasoning or inference. But it would be nice to know more about this process. What exactly is the nature of scientific reasoning? And how much confidence should we place in the inferences scientists make? These are the topics of this chapter.

Deduction and induction

Logicians make an important distinction between deductive and inductive patterns of reasoning. An example of a piece of deductive reasoning, or a deductive inference, is the following:

All Frenchmen like red wine
Pierre is a Frenchman

Therefore, Pierre likes red wine

The first two statements are called the premisses of the inference, while the third statement is called the conclusion. This is a deductive inference because it has the following property: if the premisses are true, then the conclusion must be true too. In other words, if it's true that all Frenchman like red wine, and if it's true that Pierre is a Frenchman, it follows that Pierre does indeed like red wine. This is sometimes expressed by saying that the premisses of the inference entail the conclusion. Of course, the premisses of this inference are almost certainly not true – there are bound to be Frenchmen who do not like red wine. But that is not the point. What makes the inference deductive is the existence of an appropriate relation between premisses and conclusion, namely that if the premisses are true, the conclusion must be true too. Whether the premisses are actually true is a different matter, which doesn't affect the status of the inference as deductive.

Not all inferences are deductive. Consider the following example:

The first five eggs in the box were rotten
All the eggs have the same best-before date stamped on them

Therefore, the sixth egg will be rotten too

This looks like a perfectly sensible piece of reasoning. But nonetheless it is not deductive, for the premisses do not entail the conclusion. Even if the first five eggs were indeed rotten, and even if all the eggs do have the same best-before date stamped on them, this does not guarantee that the sixth egg will be rotten too. It is quite conceivable that the sixth egg will be perfectly good. In other words, it is logically possible for the premisses of this inference to be true and yet the conclusion false, so the inference is not deductive. Instead it is known as an inductive inference. In inductive inference, or inductive reasoning, we move from premisses about objects we have examined to conclusions about objects we haven't examined – in this example, eggs.

Deductive reasoning is a much safer activity than inductive reasoning. When we reason deductively, we can be certain that if we start with true premises, we will end up with a true conclusion. But the same does not hold for inductive reasoning. On the contrary, inductive reasoning is quite capable of taking us from true premises to a false conclusion. Despite this defect, we seem to rely on inductive reasoning throughout our lives, often without even thinking about it. For example, when you turn on your computer in the morning, you are confident it will not explode in your face. Why? Because you turn on your computer every morning, and it has never exploded in your face up to now. But the inference from 'up until now, my computer has not exploded when I turned it on' to 'my computer will not explode when I turn it on this time' is inductive, not deductive. The premiss of this inference does not entail the conclusion. It is logically possible that your computer will explode this time, even though it has never done so previously.

Other examples of inductive reasoning in everyday life can readily be found. When you turn the steering wheel of your car anticlockwise, you assume the car will go to the left not the right. Whenever you drive in traffic, you effectively stake your life on this assumption. But what makes you so sure that it's true? If someone asked you to justify your conviction, what would you say? Unless you are a mechanic, you would probably reply: 'every time I've turned the steering wheel anticlockwise in the past, the car has gone to the left. Therefore, the same will happen when I turn the steering wheel anticlockwise this time.' Again, this is an inductive inference, not a deductive one. Reasoning inductively seems to be an indispensable part of everyday life.

Do scientists use inductive reasoning too? The answer seems to be yes. Consider the genetic disease known as Down's syndrome (DS for short). Geneticists tell us that DS sufferers have an additional chromosome – they have 47 instead of the normal 46 (Figure 5). How do they know this? The answer, of course, is that they

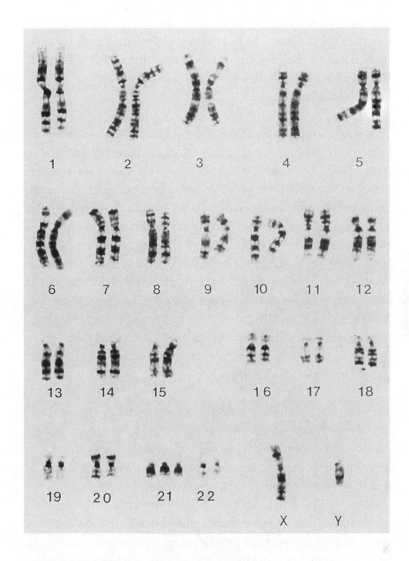

5. A representation of the complete set of chromosomes – or
karyotype – of a person with Down's syndrome. There are three copies
of chromosome 21, as opposed to the two copies most people have,
giving 47 chromosomes in total.

examined a large number of DS sufferers and found that each had an additional chromosome. They then reasoned inductively to the conclusion that all DS sufferers, including ones they hadn't examined, have an additional chromosome. It is easy to see that this inference is inductive. The fact that the DS sufferers in the sample studied had 47 chromosomes doesn't prove that all DS sufferers do. It is possible, though unlikely, that the sample was an unrepresentative one.

This example is by no means an isolated one. In effect, scientists use inductive reasoning whenever they move from limited data to a more general conclusion, which they do all the time. Consider, for example, Newton's principle of universal gravitation, encountered in the last chapter, which says that every body in the universe exerts a gravitational attraction on every other body. Now obviously, Newton did not arrive at this principle by examining every single body in the whole universe – he couldn't possibly have. Rather, he saw that the principle held true for the planets and the sun, and for objects of various sorts moving near the earth's surface. From this data, he inferred that the principle held true for all bodies. Again, this inference was obviously an inductive one: the fact that Newton's principle holds true for some bodies doesn't guarantee that it holds true for all bodies.

The central role of induction in science is sometimes obscured by the way we talk. For example, you might read a newspaper report that says that scientists have found 'experimental proof' that genetically modified maize is safe for humans. What this means is that the scientists have tested the maize on a large number of humans, and none of them have come to any harm. But strictly speaking this doesn't *prove* that the maize is safe, in the sense in which mathematicians can prove Pythagoras' theorem, say. For the inference from 'the maize didn't harm any of the people on whom it was tested' to 'the maize will not harm anyone' is inductive, not deductive. The newspaper report should really have said that scientists have found extremely good *evidence* that the maize is safe for humans. The word 'proof' should strictly only be used when we are dealing with deductive inferences. In this strict sense of the word, scientific hypotheses can rarely, if ever, be proved true by the data.

Most philosophers think it's obvious that science relies heavily on inductive reasoning, indeed so obvious that it hardly needs arguing for. But, remarkably, this was denied by the philosopher Karl Popper, who we met in the last chapter. Popper claimed that scientists only need to use deductive inferences. This would be nice if it were true, for deductive inferences are much safer than inductive ones, as we have seen.

Popper's basic argument was this. Although it is not possible to prove that a scientific theory is true from a limited data sample, it is possible to prove that a theory is false. Suppose a scientist is considering the theory that all pieces of metal conduct electricity. Even if every piece of metal she examines does conduct electricity, this doesn't prove that the theory is true, for reasons that we've seen. But if she finds even one piece of metal that does not conduct electricity, this does prove that the theory is false. For the inference from 'this piece of metal does not conduct electricity' to 'it is false that all pieces of metal conduct electricity' is a deductive inference – the premiss entails the conclusion. So if a scientist is only interested in demonstrating that a given theory is false, she may be able to accomplish her goal without the use of inductive inferences.

The weakness of Popper's argument is obvious. For scientists are not only interested in showing that certain theories are false. When a scientist collects experimental data, her aim might be to show that a particular theory – her arch-rival's theory perhaps – is false. But much more likely, she is trying to convince people that her own theory is true. And in order to do that, she will have to resort to inductive reasoning of some sort. So Popper's attempt to show that science can get by without induction does not succeed.

Hume's problem

Although inductive reasoning is not logically watertight, it nonetheless seems like a perfectly sensible way of forming beliefs about the world. The fact that the sun has risen every day up until now may not prove that it will rise tomorrow, but surely it gives us very good reason to think it will? If you came across someone who professed to be entirely agnostic about whether the sun will rise tomorrow or not, you would regard them as very strange indeed, if not irrational.

But what justifies this faith we place in induction? How should we go about persuading someone who refuses to reason inductively that they are wrong? The 18th-century Scottish philosopher David Hume (1711–1776) gave a simple but radical answer to this question. He argued that the use of induction cannot be rationally justified at all. Hume admitted that we use induction all the time, in everyday life and in science, but he insisted this was just a matter of brute animal habit. If challenged to provide a good reason for using induction, we can give no satisfactory answer, he thought.

How did Hume arrive at this startling conclusion? He began by noting that whenever we make inductive inferences, we seem to presuppose what he called the 'uniformity of nature' (UN). To see what Hume means by this, recall some of the inductive inferences from the last section. We had the inference from 'my computer hasn't exploded up to now' to 'my computer won't explode today'; from 'all examined DS sufferers have an extra chromosome' to 'all DS sufferers have an extra chromosome'; from 'all bodies observed so far obey Newton's law of gravity' to 'all bodies obey Newton's law of gravity'; and so on. In each of these cases, our reasoning seems to depend on the assumption that objects we haven't examined will be similar, in the relevant respects, to objects of the same sort that we have examined. That assumption is what Hume means by the uniformity of nature.

But how do we know that the UN assumption is actually true, Hume asks? Can we perhaps prove its truth somehow (in the strict sense of proof)? No, says Hume, we cannot. For it is easy to imagine a universe where nature is not uniform, but changes its course randomly from day to day. In such a universe, computers might sometimes explode for no reason, water might sometimes intoxicate us without warning, billiard balls might sometimes stop dead on colliding, and so on. Since such a 'non-uniform' universe is conceivable, it follows that we cannot strictly prove the truth of UN. For if we could prove that UN is true, then the non-uniform universe would be a logical impossibility.

Granted that we cannot prove UN, we might nonetheless hope to find good empirical evidence for its truth. After all, since UN has always held true up to now, surely that gives us good reason for thinking it is true? But this argument begs the question, says Hume! For it is itself an inductive argument, and so itself depends on the UN assumption. An argument that assumes UN from the outset clearly cannot be used to show that UN is true. To put the point another way, it is certainly an established fact that nature has behaved largely uniformly up to now. But we cannot appeal to this fact to argue that nature will continue to be uniform, because this assumes that what has happened in the past is a reliable guide to what will happen in the future – which *is* the uniformity of nature assumption. If we try to argue for UN on empirical grounds, we end up reasoning in a circle.

The force of Hume's point can be appreciated by imagining how you would go about persuading someone who doesn't trust inductive reasoning that they should. You would probably say: 'look, inductive reasoning has worked pretty well up until now. By using induction scientists have split the atom, landed men on the moon, invented computers, and so on. Whereas people who haven't used induction have tended to die nasty deaths. They have eaten arsenic believing that it would nourish them, jumped off tall buildings believing that they would fly, and so on (Figure 6). Therefore it will clearly pay you

6. What happens to people who don't trust induction.

to reason inductively.' But of course this wouldn't convince the doubter. For to argue that induction is trustworthy because it has worked well up to now is to reason in an inductive way. Such an argument would carry no weight with someone who doesn't already trust induction. That is Hume's fundamental point.

So the position is this. Hume points out that our inductive inferences rest on the UN assumption. But we cannot prove that UN is true, and we cannot produce empirical evidence for its truth without begging the question. So our inductive inferences rest on an assumption about the world for which we have no good grounds. Hume concludes that our confidence in induction is just blind faith – it admits of no rational justification whatever.

This intriguing argument has exerted a powerful influence on the philosophy of science, and continues to do so today. (Popper's unsuccessful attempt to show that scientists need only use deductive inferences was motivated by his belief that Hume had shown the total irrationality of inductive reasoning.) The influence of Hume's argument is not hard to understand. For normally we think of science as the very paradigm of rational enquiry. We place great faith in what scientists tell us about the world. Every time we travel by aeroplane, we put our lives in the hands of the scientists who designed the plane. But science relies on induction, and Hume's argument seems to show that induction cannot be rationally justified. If Hume is right, the foundations on which science is built do not look quite as solid as we might have hoped. This puzzling state of affairs is known as Hume's problem of induction.

Philosophers have responded to Hume's problem in literally dozens of different ways; this is still an active area of research today. Some people believe the key lies in the concept of probability. This suggestion is quite plausible. For it is natural to think that although the premises of an inductive inference do not guarantee the truth of the conclusion, they do make it quite probable. So even if scientific knowledge cannot be certain, it may nonetheless be highly probable. But this response to Hume's problem generates difficulties of its own, and is by no means universally accepted; we will return to it in due course.

Another popular response is to admit that induction cannot be rationally justified, but to argue that this is not really so problematic after all. How might one defend such a position? Some philosophers have argued that induction is so fundamental to how we think and reason that it's not the sort of thing that could be justified. Peter Strawson, an influential contemporary philosopher, defended this view with the following analogy. If someone worried about whether a particular action was legal, they could consult the law-books and compare the action with what the law-books say. But suppose someone worried about whether the law itself was legal. This is an odd worry indeed. For the law is the standard against which the legality of other things is judged, and it makes little sense to enquire whether the standard itself is legal. The same applies to induction, Strawson argued. Induction is one of the standards we use to decide whether claims about the world are justified. For example, we use induction to judge whether a pharmaceutical company's claim about the amazing benefits of its new drug are justified. So it makes little sense to ask whether induction itself is justified.

Has Strawson really succeeded in defusing Hume's problem? Some philosophers say yes, others say no. But most people agree that it is very hard to see how there could be a satisfactory justification of induction. (Frank Ramsey, a Cambridge philosopher from the 1920s, said that to ask for a justification of induction was 'to cry for the moon'.) Whether this is something that should worry us, or shake our faith in science, is a difficult question that you should ponder for yourself.

Inference to the best explanation

The inductive inferences we've examined so far have all had essentially the same structure. In each case, the premiss of the inference has had the form 'all x's examined so far have been y', and the conclusion has had the form 'the next x to be examined will be y', or sometimes, 'all x's are y'. In other words, these inferences take us from examined to unexamined instances of a given kind.

Such inferences are widely used in everyday life and in science, as we have seen. However, there is another common type of non-deductive inference that doesn't fit this simple pattern. Consider the following example:

> The cheese in the larder has disappeared, apart from a
>   few crumbs
> Scratching noises were heard coming from the larder last night
> _____
>
> Therefore, the cheese was eaten by a mouse

It is obvious that this inference is non-deductive: the premisses do not entail the conclusion. For the cheese could have been stolen by the maid, who cleverly left a few crumbs to make it look like the handiwork of a mouse (Figure 7). And the scratching noises could have been caused in any number of ways – perhaps they were due to the boiler overheating. Nonetheless, the inference is clearly a reasonable one. For the hypothesis that a mouse ate the cheese seems to provide a better explanation of the data than do the various alternative explanations. After all, maids do not normally steal cheese, and modern boilers do not tend to overheat. Whereas mice do normally eat cheese when they get the chance, and do tend to make scratching sounds. So although we cannot be certain that the mouse hypothesis is true, on balance it looks quite plausible: it is the best way of accounting for the available data.

29

7. The mouse hypothesis and the maid hypothesis can both account for the missing cheese.

Reasoning of this sort is known as 'inference to the best explanation', for obvious reasons, or IBE for short. Certain terminological confusions surround the relation between IBE and induction. Some philosophers describe IBE as a type of inductive inference; in effect, they use 'inductive inference' to mean 'any inference which is not deductive'. Others contrast IBE with inductive inference, as we have done above. On this way of cutting the pie, 'inductive inference' is reserved for inferences from examined to unexamined instances of a given kind, of the sort we examined earlier; IBE and inductive inference are then two different types of non-deductive inference. Nothing hangs on which choice of terminology we favour, so long as we stick to it consistently.

Scientists frequently use IBE. For example, Darwin argued for his theory of evolution by calling attention to various facts about the living world which are hard to explain if we assume that current species have been separately created, but which make perfect sense if current species have descended from common ancestors, as his theory held. For example, there are close anatomical similarities between the legs of horses and zebras. How do we explain this, if God created horses and zebras separately? Presumably he could have made their legs as different as he pleased. But if horses and zebras have both descended from a recent common ancestor, this provides an obvious explanation of their anatomical similarity. Darwin argued that the ability of his theory to explain facts of this sort, and of many other sorts too, constituted strong evidence for its truth.

Another example of IBE is Einstein's famous work on Brownian motion. Brownian motion refers to the chaotic, zig-zag motion of microscopic particles suspended in a liquid or gas. It was discovered in 1827 by the Scottish botanist Robert Brown (1713–1858), while examining pollen grains floating in water. A number of attempted explanations of Brownian motion were advanced in the 19th century. One theory attributed the motion to electrical attraction between particles, another to agitation from external surroundings, and another to convection currents in the fluid. The correct explanation is based on the kinetic theory of matter, which says that liquids and gases are made up of atoms or molecules in motion. The suspended particles collide with the surrounding molecules, causing the erratic, random movements that Brown first observed. This theory was first proposed in the late 19th century but was not widely accepted, not least because many scientists didn't believe that atoms and molecules were real physical entities. But in 1905, Einstein provided an ingenious mathematical treatment of

Brownian motion, making a number of precise, quantitative predictions which were later confirmed experimentally. After Einstein's work, the kinetic theory was quickly agreed to provide a far better explanation of Brownian motion than any of the alternatives, and scepticism about the existence of atoms and molecules rapidly subsided.

One interesting question is whether IBE or ordinary induction is a more fundamental pattern of inference. The philosopher Gilbert Harman has argued that IBE is more fundamental. According to this view, whenever we make an ordinary inductive inference such as 'all pieces of metal examined so far conduct electricity, therefore all pieces of metal conduct electricity' we are implicitly appealing to explanatory considerations. We assume that the correct explanation for why the pieces of metal in our sample conducted electricity, whatever it is, entails that all pieces of metal will conduct electricity; that is why we make the inductive inference. But if we believed, for example, that the explanation for why the pieces of metal in our sample conducted electricity was that a laboratory technician had tinkered with them, we would not infer that all pieces of metal conduct electricity. Proponents of this view do not say there is no difference between IBE and ordinary induction – there clearly is. Rather, they think that ordinary induction is ultimately dependent on IBE.

However, other philosophers argue that this gets things backwards: IBE is itself parasitic on ordinary induction, they say. To see the grounds for this view, think back to the cheese-in-the-larder example above. Why do we regard the mouse hypothesis as a better explanation of the data than the maid hypothesis? Presumably, because we know that maids do not normally steal cheese, whereas mice do. But this is knowledge that we have gained through ordinary inductive reasoning, based on our previous observations of the behaviour of mice and maids. So according to this view, when we try to decide which of a group of competing hypotheses provides the best explanation of our data, we invariably appeal to knowledge that has been gained through ordinary induction. Thus it is incorrect to regard IBE as a more fundamental mode of inference.

Whichever of these opposing views we favour, one issue clearly demands more attention. If we want to use IBE, we need some way of deciding which of the competing hypotheses provides the best explanation of the data. But what criteria determine this? A popular answer is that the best explanation is the simplest or the most parsimonious one. Consider again the cheese-in-the-larder example. There are two pieces of data that need explaining: the missing cheese and the scratching noises. The mouse hypothesis postulates just one cause – a mouse – to explain both pieces of data. But the maid hypothesis must postulate two causes – a dishonest maid and an overheating boiler – to explain the same data. So the mouse hypothesis is more parsimonious, hence better. Similarly in the Darwin example. Darwin's theory could explain a very diverse range of facts about the living world, not just anatomical similarities between species. Each of these facts could be explained in other ways, as Darwin knew. But the theory of evolution explained all the facts in one go – that is what made it the best explanation of the data.

The idea that simplicity or parsimony is the mark of a good explanation is quite appealing, and certainly helps flesh out the idea of IBE. But if scientists use simplicity as a guide to inference, this raises a problem. For how do we know that the universe is simple rather than complex? Preferring a theory that explains the data in terms of the fewest number of causes does seem sensible. But is there any objective reason for thinking that such a theory is more likely to be true than a less simple theory? Philosophers of science do not agree on the answer to this difficult question.

Probability and induction

The concept of probability is philosophically puzzling. Part of the puzzle is that the word 'probability' seems to have more than one meaning. If you read that the probability of an Englishwoman living to 100 years of age is 1 in 10, you would understand this as saying that one-tenth of all Englishwomen live to the age of 100. Similarly, if you read that the probability of a male smoker developing lung cancer is 1 in 4, you would take this to mean that a quarter of all male smokers develop lung cancer. This is known as the frequency interpretation of probability: it equates probabilities with proportions, or frequencies. But what if you read that the probability of finding life on Mars is 1 in 1,000? Does this mean that one out of every thousand planets in our solar system contains life? Clearly it does not. For one thing, there are only nine planets in our solar system. So a different notion of probability must be at work here.

One interpretation of the statement 'the probability of life on Mars is 1 in 1,000' is that the person who utters it is simply reporting a subjective fact about themselves – they are telling us how likely they think life on Mars is. This is the subjective interpretation of probability. It takes probability to be a measure of the strength of our personal opinions. Clearly, we hold some of our opinions more strongly than others. I am very confident that Brazil will win the World Cup, reasonably confident that Jesus Christ existed, and rather less confident that global environmental disaster can be averted. This could be expressed by saying that I assign a high probability to the statement 'Brazil will win the World Cup', a fairly high probability to 'Jesus Christ existed', and a low probability to 'global environmental disaster can be averted'. Of course, to put an exact number on the strength of my conviction in these statements would be hard, but advocates of the subjective interpretation regard this as a merely practical limitation. In principle, we should be able to assign a precise numerical probability to each of the statements about which we have an opinion, reflecting how strongly we believe or disbelieve them, they say.

The subjective interpretation of probability implies that there are no objective facts about probability, independently of what people believe. If I say that the probability of finding life on Mars is high and you say that it is very low, neither of us is right or wrong – we are both simply stating how strongly we believe the statement in question. Of course, there is an objective fact about whether there is life on Mars or not; there is just no objective fact about how probable it is that there is life on Mars, according to the subjective interpretation.

The logical interpretation of probability rejects this position. It holds that a statement such as 'the probability of life on Mars is high' is objectively true or false, relative to a specified body of evidence. A statement's probability is the measure of the strength of evidence in its favour, on this view. Advocates of the logical interpretation think that for any two statements in our language, we can in principle discover the probability of one, given the other as evidence. For example, we might want to discover the probability that there will be an ice age within 10,000 years, given the current rate of global warming. The subjective interpretation says there is no objective fact about this probability. But the logical interpretation insists that there is: the current rate of global warming confers a definite numerical probability on the occurrence of an ice age within 10,000 years, say 0.9 for example. A probability of 0.9 clearly counts as a high probability – for the maximum is 1 – so the statement 'the probability that there will be an ice age within 10,000 years is high' would then be objectively true, given the evidence about global warming.

If you have studied probability or statistics, you may be puzzled by this talk of different interpretations of probability. How do these interpretations tie in with what you learned? The answer is that the mathematical study of probability does not by itself tell us what probability means, which is what we have been examining above. Most statisticians would in fact favour the frequency interpretation, but the problem of how to interpret probability, like most philosophical problems, cannot be resolved mathematically. The mathematical formulae for working out probabilities remain the same, whichever interpretation we adopt.

Philosophers of science are interested in probability for two main reasons. The first is that in many branches of science, especially physics and biology, we find laws and theories that are formulated using the notion of probability. Consider, for example, the theory known as Mendelian genetics, which deals with the transmission of genes from one generation to another in sexually reproducing populations. One of the most important principles of Mendelian genetics is that every gene in an organism has a 50% chance of making it into any one of the organism's gametes (sperm or egg cells). Hence there is a 50% chance that any gene found in your mother will also be in you, and likewise for the genes in your father. Using this principle and others, geneticists can provide detailed explanations for why particular characteristics (e.g. eye colour) are distributed across the generations of a family in the way that they are. Now 'chance' is just another word for probability, so it is obvious that our Mendelian principle makes essential use of the concept of probability. Many other examples could be given of scientific laws and principles that are expressed in terms of probability. The need to understand these laws and principles is an important motivation for the philosophical study of probability.

The second reason why philosophers of science are interested in the concept of probability is the hope that it might shed some light on inductive inference, in particular on Hume's problem; this shall be our focus here. At the root of Hume's problem is the fact that the premises of an inductive inference do not guarantee the truth of its conclusion. But it is tempting to suggest that the premises of a typical inductive inference do make the conclusion highly probable. Although the fact that all objects examined so far obey Newton's law of gravity doesn't prove that all objects do, surely it does make it very probable? So surely Hume's problem can be answered quite easily after all?

However, matters are not quite so simple. For we must ask what interpretation of probability this response to Hume assumes. On the frequency interpretation, to say it is highly probable that all objects obey Newton's law is to say that a very high proportion of all objects obey the law. But there is no way we can know that, unless we use induction! For we have only examined a tiny fraction of all the objects in the universe. So Hume's problem remains. Another way to see the point is this. We began with the inference from 'all examined objects obey Newton's law' to 'all objects obey Newton's law'. In response to Hume's worry that the premiss of this inference doesn't guarantee the truth of the conclusion, we suggested that it might nonetheless make the conclusion highly probable. But the inference from 'all examined objects obey Newton's law' to 'it is highly probable that all objects obey Newton's law' is still an inductive inference, given that the latter means 'a very high proportion of all objects obey Newton's law', as it does according to the frequency interpretation. So appealing to the concept of probability does not take the sting out of Hume's argument, if we adopt a frequency interpretation of probability. For knowledge of probabilities then becomes itself dependent on induction.

The subjective interpretation of probability is also powerless to solve Hume's problem, though for a different reason. Suppose John believes that the sun will rise tomorrow and Jack believes it will not. They both accept the evidence that the sun has risen every day in the past. Intuitively, we want to say that John is rational and Jack isn't, because the evidence makes John's belief more probable. But if probability is simply a matter of subjective opinion, we cannot say this. All we can say is that John assigns a high probability to 'the sun will rise tomorrow' and Jack does not. If there are no objective facts about probability, then we cannot say that the conclusions of inductive inferences are objectively probable. So we have no explanation of why someone like Jack, who declines to use induction, is irrational. But Hume's problem is precisely the demand for such an explanation.

The logical interpretation of probability holds more promise of a satisfactory response to Hume. Suppose there is an objective fact about the probability that the sun will rise tomorrow, given that it has risen every day in the past. Suppose this probability is very high. Then we have an explanation of why John is rational and Jack isn't. For John and Jack both accept the evidence that the sun has risen every day in the past, but Jack fails to realize that this evidence makes it highly probable that the sun will rise tomorrow, while John does realize this. Regarding a statement's probability as a measure of the evidence in its favour, as the logical interpretation recommends, tallies neatly with our intuitive feeling that the premises of an inductive inference can make the conclusion highly probable, even if they cannot guarantee its truth.

Unsurprisingly, therefore, those philosophers who have tried to solve Hume's problem via the concept of probability have tended to favour the logical interpretation. (One of these was the famous economist John Maynard Keynes, whose early interests were in logic and philosophy.) Unfortunately, most people today believe that the logical interpretation of probability faces very serious, probably insuperable, difficulties. This is because all the attempts to work out the logical interpretation of probability in any detail have run up against a host of problems, both mathematical and philosophical. As a result, many philosophers today are inclined to reject outright the underlying assumption of the logical interpretation – that there are objective facts about the probability of one statement, given another. Rejecting this assumption leads naturally to the subjective interpretation of probability, but that, as we have seen, offers scant hope of a satisfactory response to Hume.

Even if Hume's problem is ultimately insoluble, as seems likely, thinking about the problem is still a valuable exercise. For reflecting on the problem of induction leads us into a thicket of interesting questions about the structure of scientific reasoning, the nature of rationality, the appropriate degree of confidence to place in science, the interpretation of probability, and more. Like most philosophical questions, these questions probably do not admit of final answers, but in grappling with them we learn much about the nature and limits of scientific knowledge.

Chapter 3
Explanation in science

One of the most important aims of science is to try and explain what happens in the world around us. Sometimes we seek explanations for practical ends. For example, we might want to know why the ozone layer is being depleted so quickly, in order to try and do something about it. In other cases we seek scientific explanations simply to satisfy our intellectual curiosity – we want to understand more about how the world works. Historically, the pursuit of scientific explanation has been motivated by both goals.

Quite often, modern science is successful in its aim of supplying explanations. For example, chemists can explain why sodium turns yellow when it burns. Astronomers can explain why solar eclipses occur when they do. Economists can explain why the yen declined in value in the 1980s. Geneticists can explain why male baldness tends to run in families. Neurophysiologists can explain why extreme oxygen deprivation leads to brain damage. You can probably think of many other examples of successful scientific explanations.

But what exactly *is* scientific explanation? What exactly does it mean to say that a phenomenon can be 'explained' by science? This is a question that has exercised philosophers since Aristotle, but our starting point will be a famous account of scientific explanation put forward in the 1950s by the American philosopher Carl Hempel.

Hempel's account is known as the *covering law* model of explanation, for reasons that will become clear.

Hempel's covering law model of explanation

The basic idea behind the covering law model is straightforward. Hempel noted that scientific explanations are usually given in response to what he called 'explanation-seeking why questions'. These are questions such as 'why is the earth not perfectly spherical?', 'why do women live longer than men?', and the like – they are demands for explanation. To give a scientific explanation is thus to provide a satisfactory answer to an explanation-seeking why question. If we could determine the essential features that such an answer must have, we would know what scientific explanation is.

Hempel suggested that scientific explanations typically have the logical structure of an argument, i.e. a set of premises followed by a conclusion. The conclusion states that the phenomenon that needs explaining actually occurs, and the premises tell us why the conclusion is true. Thus suppose someone asks why sugar dissolves in water. This is an explanation-seeking why question. To answer it, says Hempel, we must construct an argument whose conclusion is 'sugar dissolves in water' and whose premises tell us why this conclusion is true. The task of providing an account of scientific explanation then becomes the task of characterizing precisely the relation that must hold between a set of premises and a conclusion, in order for the former to count as an explanation of the latter. That was the problem Hempel set himself.

Hempel's answer to the problem was three-fold. Firstly, the premises should entail the conclusion, i.e. the argument should be a deductive one. Secondly, the premises should all be true. Thirdly, the premises should consist of at least one general law. General laws are things such as 'all metals conduct electricity', 'a body's acceleration varies inversely with its mass', 'all plants contain chlorophyll', and so on; they contrast with particular facts such as 'this piece of metal conducts electricity', 'the plant on my desk contains chlorophyll' and so on. General laws are sometimes called 'laws of nature'. Hempel allowed that a scientific explanation could appeal to particular facts as well as general laws, but he held that at least one general law was always essential. So to explain a phenomenon, on Hempel's conception, is to show that its occurrence follows deductively from a general law, perhaps supplemented by other laws and/or particular facts, all of which must be true.

To illustrate, suppose I am trying to explain why the plant on my desk has died. I might offer the following explanation. Owing to the poor light in my study, no sunlight has been reaching the plant; but sunlight is necessary for a plant to photosynthesize; and without photosynthesis a plant cannot make the carbohydrates it needs to survive, and so will die; therefore my plant died. This explanation fits Hempel's model exactly. It explains the death of the plant by deducing it from two true laws – that sunlight is necessary for photosynthesis, and that photosynthesis is necessary for survival – and one particular fact – that the plant was not getting any sunlight. Given the truth of the two laws and the particular fact, the death of the plant *had* to occur; that is why the former constitute a good explanation of the latter.

Schematically, Hempel's model of explanation can be written as follows:

General laws
Particular facts
$\Rightarrow$
Phenomenon to be explained

The phenomenon to be explained is called the *explanandum*, and the general laws and particular facts that do the explaining are called the *explanans*. The explanandum itself may be either a particular fact or a general law. In the example above, it was a particular fact – the death of my plant. But sometimes the things we want to explain are general. For example, we might wish to explain why exposure to the sun leads to skin cancer. This is a general law, not a particular fact. To explain it, we would need to deduce it from still more fundamental laws – presumably, laws about the impact of radiation on skin cells, combined with particular facts about the amount of radiation in sunlight. So the structure of a scientific explanation is essentially the same, whether the *explanandum*, i.e. the thing we are trying to explain, is particular or general.

It is easy to see why Hempel's model is called the covering law model of explanation. For according to the model, the essence of explanation is to show that the phenomenon to be explained is 'covered' by some general law of nature. There is certainly something appealing about this idea. For showing that a phenomenon is a consequence of a general law does in a sense take the mystery out of it – it renders it more intelligible. And in fact, scientific explanations do often fit the pattern Hempel describes. For example, Newton explained why the planets move in ellipses around the sun by showing that this can be deduced from his law of universal gravitation, along with some minor additional assumptions. Newton's explanation fits Hempel's model exactly: a phenomenon is explained by showing that it had to be so, given the laws of nature plus some additional facts. After Newton, there was no longer any mystery about why planetary orbits are elliptical.

Hempel was aware that not all scientific explanations fit his model exactly. For example, if you ask someone why Athens is always immersed in smog, they will probably say 'because of car exhaust pollution'. This is a perfectly acceptable scientific explanation, though it involves no mention of any laws. But Hempel would say that if the explanation were spelled out in full detail, laws would enter the picture. Presumably there is a law that says something like 'if carbon monoxide is released into the earth's atmosphere in sufficient concentration, smog clouds will form'. The full explanation of why Athens is bathed in smog would cite this law, along with the fact that car exhaust contains carbon monoxide and

Athens has lots of cars. In practice, we wouldn't spell out the explanation in this much detail unless we were being very pedantic. But if we were to spell it out, it would correspond quite well to the covering law pattern.

Hempel drew an interesting philosophical consequence from his model about the relation between explanation and prediction. He argued that these are two sides of the same coin. Whenever we give a covering law explanation of a phenomenon, the laws and particular facts we cite would have enabled us to predict the occurrence of the phenomenon, if we hadn't already known about it. To illustrate, consider again Newton's explanation of why planetary orbits are elliptical. This fact was known long before Newton explained it using his theory of gravity – it was discovered by Kepler. But if it had not been known, Newton would have been able to predict it from his theory of gravity, for his theory entails that planetary orbits are elliptical, given minor additional assumptions. Hempel expressed this by saying that every scientific explanation is potentially a prediction – it would have served to predict the phenomenon in question, had it not already been known. The converse was also true, Hempel thought: every reliable prediction is potentially an explanation. To illustrate, suppose scientists predict that mountain gorillas will be extinct by 2010, based on information about the destruction of their habitat. Suppose they turn out to be right. According to Hempel, the information they used to predict the gorillas' extinction before it happened will serve to explain that same fact after it has happened. Explanation and prediction are structurally symmetric.

Though the covering law model captures the structure of many actual scientific explanations quite well, it also faces a number of awkward counter-examples. These counter-examples fall into two classes. On the one hand, there are cases of genuine scientific explanations that do not fit the covering law model, even approximately. These cases suggest that Hempel's model is too strict – it excludes some *bona fide* scientific explanations. On the

44

other hand, there are cases of things that *do* fit the covering law model, but intuitively do not count as genuine scientific explanations. These cases suggest that Hempel's model is too liberal – it allows in things that should be excluded. We will focus on counter-examples of the second sort.

The problem of symmetry

Suppose you are lying on the beach on a sunny day, and you notice that a flagpole is casting a shadow of 20 metres across the sand (Figure 8).

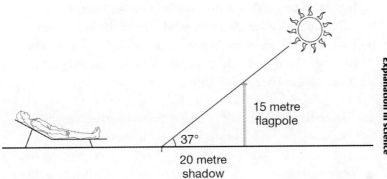

8. A 15-metre flagpole casts a shadow of 20 metres on the beach when the sun is 37° overhead.

Someone asks you to explain why the shadow is 20 metres long. This is an explanation-seeking why question. A plausible answer might go as follows: 'light rays from the sun are hitting the flagpole, which is exactly 15 metres high. The angle of elevation of the sun is 37°. Since light travels in straight lines, a simple trigonometric calculation (tan 37° = 15/20) shows that the flagpole will cast a shadow 20 metres long'.

This looks like a perfectly good scientific explanation. And by rewriting it in accordance with Hempel's schema, we can see that it fits the covering law model:

45

General laws	Light travels in straight lines
	Laws of trigonometry
Particular facts	Angle of elevation of the sun is 37°
	Flagpole is 15 metres high
⇒	
Phenomenon to be explained	Shadow is 20 metres long

The length of the shadow is deduced from the height of the flagpole and the angle of elevation of the sun, along with the optical law that light travels in straight lines and the laws of trigonometry. Since these laws are true, and since the flagpole is indeed 15 metres high, the explanation satisfies Hempel's requirements precisely. So far so good. The problem arises as follows. Suppose we swap the *explanandum* – that the shadow is 20 metres long – with the particular fact that the flagpole is 15 metres high. The result is this:

General law	Light travels in straight lines
	Laws of trigonometry
Particular facts	Angle of elevation of the sun is 37°
	Shadow is 20 metres long
⇒	
Phenomenon to be explained	Flagpole is 15 metres high

This 'explanation' clearly conforms to the covering law pattern too. The height of the flagpole is deduced from the length of the shadow it casts and the angle of elevation of the sun, along with the optical law that light travels in straight lines and the laws of trigonometry. But it seems very odd to regard this as an *explanation* of why the flagpole is 15 metres high. The real explanation of why the flagpole is 15 metres high is presumably that a carpenter deliberately made it so – it has nothing to do with the length of the shadow that it casts. So Hempel's model is too liberal: it allows something to count as a scientific explanation that obviously is not.

The general moral of the flagpole example is that the concept of explanation exhibits an important asymmetry. The height of the flagpole explains the length of the shadow, given the relevant laws and additional facts, but not vice-versa. In general, if x explains y, given the relevant laws and additional facts, then it will not be true that y explains x, given the same laws and facts. This is sometimes expressed by saying that explanation is an asymmetric relation. Hempel's covering law model does not respect this asymmetry. For just as we can deduce the length of the shadow from the height of the flagpole, given the laws and additional facts, so we can deduce the height of the flagpole from the length of the shadow. In other words, the covering law model implies that explanation should be a symmetric relation, but in fact it is asymmetric. So Hempel's model fails to capture fully what it is to be a scientific explanation.

The shadow and flagpole case also provides a counter-example to Hempel's thesis that explanation and prediction are two sides of the same coin. The reason is obvious. Suppose you didn't know how high the flagpole was. If someone told you that it was casting a shadow of 20 metres and that the sun was 37° overhead, you would be able to *predict* the flagpole's height, given that you knew the relevant optical and trigonometrical laws. But as we have just seen, this information clearly doesn't *explain* why the flagpole has the height it does. So in this example prediction and explanation part ways. Information that serves to predict a fact before we know it does not serve to explain that same fact after we know it, which contradicts Hempel's thesis.

The problem of irrelevance

Suppose a young child is in a hospital in a room full of pregnant women. The child notices that one person in the room – who is a man called John – is not pregnant, and asks the doctor why not. The doctor replies: 'John has been taking birth-control pills regularly for the last few years. People who take birth-control pills regularly never become pregnant. Therefore, John has not become pregnant'.

Let us suppose for the sake of argument that what the doctor says is true – John is mentally ill and does indeed take birth-control pills, which he believes help him. Even so, the doctor's reply to the child is clearly not very helpful. The correct explanation of why John has not become pregnant, obviously, is that he is male and males cannot become pregnant.

However, the explanation the doctor has given the child fits the covering law model perfectly. The doctor deduces the phenomenon to be explained – that John is not pregnant – from the general law that people who take birth-control pills do not become pregnant and the particular fact that John has been taking birth-control pills. Since both the general law and the particular fact are true, and since they do indeed entail the *explanandum*, according to the covering law model the doctor has given a perfectly adequate explanation of why John is not pregnant. But of course he hasn't. Hence the covering law model is again too permissive: it allows things to count as scientific explanations that intuitively are not.

The general moral is that a good explanation of a phenomenon should contain information that is *relevant* to the phenomenon's occurrence. This is where the doctor's reply to the child goes wrong. Although what the doctor tells the child is perfectly true, the fact that John has been taking birth-control pills is irrelevant to his not being pregnant, because he wouldn't have been pregnant even if he hadn't been taking the pills. This is why the doctor's reply does not constitute a good answer to the child's question. Hempel's model does not respect this crucial feature of our concept of explanation.

Explanation and causality

Since the covering law model encounters so many problems, it is natural to look for an alternative way of understanding scientific explanation. Some philosophers believe that the key lies in the concept of causality. This is quite an attractive suggestion. For in many cases to explain a phenomenon is indeed to say what caused it. For example, if an accident investigator is trying to explain an aeroplane crash, he is obviously looking for the cause of the crash. Indeed, the questions 'why did the plane crash?' and 'what was the cause of the plane crash?' are practically synonymous. Similarly, if an ecologist is trying to explain why there is less biodiversity in the tropical rainforests than there used to be, he is clearly looking for the cause of the reduction in biodiversity. The link between the concepts of explanation and causality is quite intimate.

Impressed by this link, a number of philosophers have abandoned the covering law account of explanation in favour of causality-based accounts. The details vary, but the basic idea behind these accounts is that to explain a phenomenon is simply to say what caused it. In some cases, the difference between the covering law and causal accounts is not actually very great, for to deduce the occurrence of a phenomenon from a general law often just *is* to give its cause. For example, recall again Newton's explanation of why planetary orbits are elliptical. We saw that this explanation fits the covering law model – for Newton deduced the shape of the planetary orbits from his law of gravity, plus some additional facts. But Newton's explanation was also a causal one, since elliptical planetary orbits are caused by the gravitational attraction between planets and the sun.

However, the covering law and causal accounts are not fully equivalent – in some cases they diverge. Indeed, many philosophers favour a causal account of explanation precisely because they think it can avoid some of the problems facing the covering law model. Recall the flagpole problem. Why do our intuitions tell us that the height of the flagpole explains the length of the shadow, given the laws, but not vice-versa? Plausibly, because the height of the flagpole is the cause of the shadow being 20 metres long, but the shadow being 20 metres long is not the cause of the flagpole being 15 metres high. So unlike the covering law model, a causal account of explanation gives the 'right' answer in the flagpole case – it respects our intuition that we cannot explain the height of the flagpole by pointing to the length of the shadow it casts.

The general moral of the flagpole problem was that the covering law model cannot accommodate the fact that explanation is an asymmetric relation. Now causality is obviously an asymmetric relation too: if x is the cause of y, then y is not the cause of x. For example, if the short-circuit caused the fire, then the fire clearly did not cause the short-circuit. It is therefore quite plausible to suggest that the asymmetry of explanation derives from the asymmetry of causality. If to explain a phenomenon is to say what caused it, then since causality is asymmetric we should expect explanation to be asymmetric too – as it is. The covering law model runs up against the flagpole problem precisely because it tries to analyse the concept of scientific explanation without reference to causality.

The same is true of the birth-control pill case. That John takes birth-control pills does not explain why he isn't pregnant, because the birth-control pills are not the cause of his not being pregnant. Rather, John's gender is the cause of his not being pregnant. That is why we think that the correct answer to the question 'why is John not pregnant?' is 'because he is a man, and men can't become pregnant', rather than the doctor's answer. The doctor's answer satisfies the covering law model, but since it does not correctly identify the cause of the phenomenon we wish to explain, it does not constitute a genuine explanation. The general moral we drew from the birth-control pill example was that a genuine scientific explanation must contain information that is relevant to the *explanandum*. In effect, this is another way of saying that the explanation should tell us the *explanandum*'s cause. Causality-based accounts of scientific explanation do not run up against the problem of irrelevance.

It is easy to criticize Hempel for failing to respect the close link between causality and explanation, and many people have done so.

In some ways, this criticism is a bit unfair. For Hempel subscribed to a philosophical doctrine known as *empiricism*, and empiricists are traditionally very suspicious of the concept of causality. Empiricism says that all our knowledge comes from experience. David Hume, whom we met in the last chapter, was a leading empiricist, and he argued that it is impossible to experience causal relations. So he concluded that they don't exist – causality is a figment of our imagination! This is a very hard conclusion to accept. Surely it is an objective fact that dropping glass vases causes them to break? Hume denied this. He allowed that it is an objective fact that most glass vases that have been dropped have in fact broken. But our idea of causality includes more than this. It includes the idea of a causal link between the dropping and the breaking, i.e. that the former brings about the latter. No such links are to be found in the world, according to Hume: all we see is a vase being dropped, and then it breaking a moment later. We experience no causal connection between the first event and the second. Causality is therefore a fiction.

Most empiricists have not accepted this startling conclusion outright. But as a result of Hume's work, they have tended to regard causality as a concept to be treated with great caution. So to an empiricist, the idea of analysing the concept of explanation in terms of the concept of causality would seem perverse. If one's goal is to clarify the concept of scientific explanation, as Hempel's was, there is little point in using notions that are equally in need of clarification themselves. And for empiricists, causality is definitely in need of philosophical clarification. So the fact that the covering law model makes no mention of causality was not a mere oversight on Hempel's part. In recent years, empiricism has declined somewhat in popularity. Furthermore, many philosophers have come to the conclusion that the concept of causality, although philosophically problematic, is indispensable to how we understand the world. So the idea of a causality-based account of scientific explanation seems more acceptable than it would have done in Hempel's day.

Causality-based accounts of explanation certainly capture the structure of many actual scientific explanations quite well, but are they the whole story? Many philosophers say no, on the grounds that certain scientific explanations do not seem to be causal. One type of example stems from what are called 'theoretical identifications' in science. Theoretical identifications involve identifying one concept with another, usually drawn from a different branch of science. 'Water is H_2O' is an example, as is 'temperature is average molecular kinetic energy'. In both of these cases, a familiar everyday concept is equated or identified with a more esoteric scientific concept. Often, theoretical identifications furnish us with what seem to be scientific explanations. When chemists discovered that water is H_2O, they thereby explained what water is. Similarly, when physicists discovered that an object's temperature is the average kinetic energy of its molecules, they thereby explained what temperature is. But neither of these explanations is causal. Being made of H_2O doesn't *cause* a substance to be water – it just *is* being water. Having a particular average molecular kinetic energy doesn't *cause* a liquid to have the temperature it does – it just *is* having that temperature. If these examples are accepted as legitimate scientific explanations, they suggest that causality-based accounts of explanation cannot be the whole story.

Can science explain everything?

Modern science can explain a great deal about the world we live in. But there are also numerous facts that have not been explained by science, or at least not explained fully. The origin of life is one such example. We know that about 4 billion years ago, molecules with the ability to make copies of themselves appeared in the primeval soup, and life evolved from there. But we do not understand how these self-replicating molecules got there in the first place. Another example is the fact that autistic children tend to have very good memories. Numerous studies of autistic children have confirmed this fact, but as yet nobody has succeeded in explaining it.

Many people believe that in the end, science will be able to explain facts of this sort. This is quite a plausible view. Molecular biologists are working hard on the problem of the origin of life, and only a pessimist would say they will never solve it. Admittedly, the problem is not easy, not least because it is very hard to know what conditions on earth 4 billion years ago were like. But nonetheless, there is no reason to think that the origin of life will never be explained. Similarly for the exceptional memories of autistic children. The science of memory is still in its infancy, and much remains to be discovered about the neurological basis of autism. Obviously we cannot guarantee that the explanation will eventually be found. But given the number of explanatory successes that modern science has already notched up, the smart money must be on many of today's unexplained facts eventually being explained too.

But does this mean that science can in principle explain everything? Or are there some phenomena that must forever elude scientific explanation? This is not an easy question to answer. On the one hand, it seems arrogant to assert that science can explain everything. On the other hand, it seems short-sighted to assert that any particular phenomenon can never be explained scientifically. For science changes and develops very fast, and a phenomenon that looks completely inexplicable from the vantage-point of today's science may be easily explained tomorrow.

According to some philosophers, there is a purely logical reason why science will never be able to explain everything. For in order to explain something, whatever it is, we need to invoke something else. But what explains the second thing? To illustrate, recall that Newton explained a diverse range of phenomena using his law of gravity. But what explains the law of gravity itself? If someone asks *why* all bodies exert a gravitational force on each other, what should we tell them? Newton had no answer to this question. In Newtonian science the law of gravity was a fundamental principle: it explained other things, but could not itself be explained. The moral is generalizable. However much the science of the future can explain, the explanations it gives will have to make use of certain fundamental laws and principles. Since nothing can explain itself, it follows that at least some of these laws and principles will themselves remain unexplained.

Whatever one makes of this argument, it is undeniably very abstract. It purports to show that some things will never be explained, but does not tell us what they are. However, some philosophers have made concrete suggestions about phenomena that they think science can never explain. An example is consciousness – the distinguishing feature of thinking, feeling creatures such as ourselves and other higher animals. Much research into the nature of consciousness has been and continues to be done, by brain scientists, psychologists, and others. But a number of recent philosophers claim that whatever this research throws up, it will never fully explain the nature of consciousness. There is something intrinsically mysterious about the phenomenon of consciousness, they maintain, that no amount of scientific investigation can eliminate.

What are the grounds for this view? The basic argument is that conscious experiences are fundamentally unlike anything else in the world, in that they have a 'subjective aspect'. Consider, for example, the experience of watching a terrifying horror movie. This is an experience with a very distinctive 'feel' to it; in the current jargon, there is 'something that it is like' to have the experience. Neuroscientists may one day be able to give a detailed account of the complex goings-on in the brain that produce our feeling of terror. But will this explain why watching a horror movie feels the way it does, rather than feeling some other way? Many people believe that it will not. On this view, the scientific study of the brain can at most tell us which brain processes are correlated with which conscious experiences. This is certainly interesting and valuable information. However, it doesn't tell us *why* experiences with distinctive subjective 'feels' should result from the purely physical goings-on in the brain. Hence consciousness, or at least one important aspect of it, is scientifically inexplicable.

Though quite compelling, this argument is very controversial and not endorsed by all philosophers, let alone all neuroscientists. Indeed, a well-known book published in 1991 by the philosopher Daniel Dennett is defiantly entitled *Consciousness Explained*. Supporters of the view that consciousness is scientifically inexplicable are sometimes accused of having a lack of imagination. Even if it is true that brain science as currently practised cannot explain the subjective aspect of conscious experience, can we not imagine the emergence of a radically different type of brain science, with radically different explanatory techniques, that *does* explain why our experiences feel the way they do? There is a long tradition of philosophers trying to tell scientists what is and isn't possible, and later scientific developments have often proved the philosophers wrong. Only time will tell whether the same fate awaits those who argue that consciousness must always elude scientific explanation.

Explanation and reduction

The different scientific disciplines are designed for explaining different types of phenomena. To explain why rubber doesn't conduct electricity is a task for physics. To explain why turtles have such long lives is a task for biology. To explain why higher interest rates reduce inflation is a task for economics, and so on. In short, there is a division of labour between the different sciences: each specializes in explaining its own particular set of phenomena. This explains why the sciences are not usually in competition with one another – why biologists, for example, do not worry that physicists and economists might encroach on their turf.

Nonetheless, it is widely held that the different branches of science are not all on a par: some are more fundamental than others. Physics is usually regarded as the most fundamental science of all.

Why? Because the objects studied by the other sciences are ultimately composed of physical particles. Consider living organisms, for example. Living organisms are made up of cells, which are themselves made up of water, nucleic acids (such as DNA), proteins, sugars, and lipids (fats), all of which consist of molecules or long chains of molecules joined together. But molecules are made up of atoms, which are physical particles. So the objects biologists study are ultimately just very complex physical entities. The same applies to the other sciences, even the social sciences. Take economics, for example. Economics studies the behaviour of corporations and consumers in the market place, and the consequences of this behaviour. But consumers are human beings and corporations are made up of human beings; and human beings are living organisms, hence physical entities.

Does this mean that, in principle, physics can subsume all the higher-level sciences? Since everything is made up of physical particles, surely if we had a complete physics, which allowed us to predict perfectly the behaviour of every physical particle in the universe, all the other sciences would become superfluous? Most philosophers resist this line of thought. After all, it seems crazy to suggest that physics might one day be able to explain the things that biology and economics explain. The prospect of deducing the laws of biology and economics straight from the laws of physics looks very remote. Whatever the physics of the future looks like, it is most unlikely to be capable of predicting economic downturns. Far from being reducible to physics, sciences such as biology and economics seem largely autonomous of it.

This leads to a philosophical puzzle. How can a science that studies entities that are ultimately physical *not* be reducible to physics? Granted that the higher-level sciences are in fact autonomous of physics, how is this possible? According to some philosophers, the answer lies in the fact that the objects studied by the higher-level sciences are 'multiply realized' at the physical level. To illustrate the idea of multiple realization, imagine a collection of ashtrays. Each individual ashtray is obviously a physical entity, like everything else in the universe. But the physical composition of the ashtrays could be very different – some might be made of glass, others of aluminium, others of plastic, and so on. And they will probably differ in size, shape, and weight. There is virtually no limit on the range of different physical properties that an ashtray can have. So it is impossible to define the concept 'ashtray' in purely physical terms. We cannot find a true statement of the form 'x is an ashtray if and only if x is . . . .' where the blank is filled by an expression taken from the language of physics. This means that ashtrays are multiply realized at the physical level.

Philosophers have often invoked multiple realization to explain why psychology cannot be reduced to physics or chemistry, but in principle the explanation works for any higher-level science. Consider, for example, the biological fact that nerve cells live longer than skin cells. Cells are physical entities, so one might think that this fact will one day be explained by physics. However, cells are almost certainly multiply realized at the microphysical level. Cells are ultimately made up of atoms, but the precise arrangement of atoms will be very different in different cells. So the concept 'cell' cannot be defined in terms drawn from fundamental physics. There is no true statement of the form 'x is a cell if and only if x is . . .' where the blank is filled by an expression taken from the language of microphysics. If this is correct, it means that fundamental physics will never be able to explain why nerve cells live longer than skin cells, or indeed any other facts about cells. The vocabulary of cell biology and the vocabulary of physics do not map onto each other in the required way. Thus we have an explanation of why it is that cell biology cannot be reduced to physics, despite the fact that cells are physical entities. Not all philosophers are happy with the doctrine of multiple realization, but it does promise to provide a neat explanation of the autonomy of the higher-level sciences, both from physics and from each other.

Chapter 4
Realism and anti-realism

There is a very ancient debate in philosophy between two
opposing schools of thought called *realism* and *idealism*. Realism
holds that the physical world exists independently of human
thought and perception. Idealism denies this – it claims that the
physical world is in some way dependent on the conscious activity
of humans. To most people, realism seems more plausible than
idealism. For realism fits well with the common-sense view that
the facts about the world are 'out there' waiting to be discovered
by us, but idealism does not. Indeed, at first glance idealism can
sound plain silly. Since rocks and trees would presumably continue
to exist even if the human race died out, in what sense is their
existence dependent on human minds? In fact, the issue is a bit
more subtle than this, and continues to be discussed by
philosophers today.

Though the traditional realism/idealism issue belongs to an area of
philosophy called *metaphysics*, it has actually got nothing in
particular to do with science. Our concern in this chapter is with a
more modern debate that is specifically about science, and is in
some ways analogous to the traditional issue. The debate is between
a position known as *scientific realism* and its converse, known as
anti-realism or *instrumentalism*. From now on, we shall use the
word 'realism' to mean scientific realism, and 'realist' to mean
scientific realist.

Scientific realism and anti-realism

Like most philosophical 'isms', scientific realism comes in many different versions, so cannot be defined in a totally precise way. But the basic idea is straightforward. Realists hold that the aim of science is to provide a true description of the world. This may sound like a fairly innocuous doctrine. For surely no-one thinks science is aiming to produce a false description of the world. But that is not what anti-realists think. Rather, anti-realists hold that the aim of science is to provide a true description of a certain *part* of the world – the 'observable' part. As far as the 'unobservable' part of the world goes, it makes no odds whether what science says is true or not, according to anti-realists.

What exactly do anti-realists mean by the observable part of the world? They mean the everyday world of tables and chairs, trees and animals, test-tubes and Bunsen burners, thunderstorms and snow showers, and so on. Things such as these can be directly perceived by human beings – that is what it means to call them observable. Some branches of science deal exclusively with objects that are observable. An example is palaeontology, or the study of fossils. Fossils are readily observable – anyone with normally functioning eyesight can see them. But other sciences make claims about the unobservable region of reality. Physics is the obvious example. Physicists advance theories about atoms, electrons, quarks, leptons, and other strange particles, none of which can be observed in the normal sense of the word. Entities of this sort lie beyond the reach of the observational powers of humans.

With respect to sciences like palaeontology, realists and anti-realists do not disagree. Since fossils are observable, the realist thesis that science aims to truly describe the world and the anti-realist thesis that science aims to truly describe the observable world obviously coincide, as far as the study of fossils is concerned. But when it comes to sciences like physics, realists and anti-realists disagree. Realists say that when physicists put forward theories about electrons and quarks, they are trying to provide a true description of the subatomic world, just as paleontologists are trying to provide a true description of the world of fossils. Anti-realists disagree: they see a fundamental difference between theories in subatomic physics and in palaeontology.

What do anti-realists think physicists *are* up to when they talk about unobservable entities? Typically they claim that these entities are merely convenient fictions, introduced by physicists in order to help predict observable phenomena. To illustrate, consider the kinetic theory of gases, which says that any volume of a gas contains a large number of very small entities in motion. These entities – molecules – are unobservable. From the kinetic theory we can deduce various consequences about the observable behaviour of gases, e.g. that heating a sample of gas will cause it to expand if the pressure remains constant, which can be verified experimentally. According to anti-realists, the only purpose of positing unobservable entities in the kinetic theory is to deduce consequences of this sort. Whether or not gases really *do* contain molecules in motion doesn't matter; the point of the kinetic theory is not to truly describe the hidden facts, but just to provide a convenient way of predicting observations. We can see why anti-realism is sometimes called 'instrumentalism' – it regards scientific theories as instruments for helping us predict observational phenomena, rather than as attempts to describe the underlying nature of reality.

Since the realism/anti-realism debate concerns the aim of science, one might think it could be resolved by simply asking the scientists themselves. Why not do a straw poll of scientists asking them about their aims? But this suggestion misses the point – it takes the expression 'the aim of science' too literally. When we ask what the aim of science is, we are not asking about the aims of individual scientists. Rather, we are asking how best to make sense of what scientists say and do – how to interpret the scientific enterprise. Realists think we should interpret all scientific theories as attempted descriptions of reality; anti-realists think this interpretation is inappropriate for theories that talk about unobservable entities and processes. While it would certainly be interesting to discover scientists' own views on the realism/anti-realism debate, the issue is ultimately a philosophical one.

Much of the motivation for anti-realism stems from the belief that we cannot actually attain knowledge of the unobservable part of reality – it lies beyond human ken. On this view, the limits to scientific knowledge are set by our powers of observation. So science can give us knowledge of fossils, trees, and sugar crystals, but not of atoms, electrons, and quarks – for the latter are unobservable. This view is not altogether implausible. For no-one could seriously doubt the existence of fossils and trees, but the same is not true of atoms and electrons. As we saw in the last chapter, in the late 19th century many leading scientists did doubt the existence of atoms. Anyone who accepts such a view must obviously give some explanation of *why* scientists advance theories about unobservable entities, if scientific knowledge is limited to what can be observed. The explanation anti-realists give is that they are convenient fictions, designed to help predict the behaviour of things in the observable world.

Realists do not agree that scientific knowledge is limited by our powers of observation. On the contrary, they believe we already have substantial knowledge of unobservable reality. For there is every reason to believe that our best scientific theories are true, and our best scientific theories talk about unobservable entities. Consider, for example, the atomic theory of matter, which says that all matter is made up of atoms. The atomic theory is capable of explaining a great range of facts about the world. According to realists, that is good evidence that the theory is true, i.e. that matter really is made up of atoms that behave as the theory says. Of course the theory *might* be false, despite the apparent evidence in its favour, but so might any theory. Just because atoms are unobservable, that is no reason to interpret atomic theory as anything other than an attempted description of reality – and a very successful one, in all likelihood.

Strictly we should distinguish two sorts of anti-realism. According to the first sort, talk of unobservable entities is not to be understood literally at all. So when a scientist puts forward a theory about electrons, for example, we should not take him to be asserting the existence of entities called 'electrons'. Rather, his talk of electrons is metaphorical. This form of anti-realism was popular in the first half of the 20th century, but few people advocate it today. It was motivated largely by a doctrine in the philosophy of language, according to which it is not possible to make meaningful assertions about things that cannot in principle be observed, a doctrine that few contemporary philosophers accept. The second sort of anti-realism accepts that talk of unobservable entities should be taken at face value: if a theory says that electrons are negatively charged, it is true if electrons do exist and are negatively charged, but false otherwise. But we will never know which, says the anti-realist. So the correct attitude towards the claims that scientists make about unobservable reality is one of total agnosticism. They are either true or false, but we are incapable of finding out which. Most modern anti-realism is of this second sort.

The 'no miracles' argument

Many theories that posit unobservable entities are *empirically successful* – they make excellent predictions about the behaviour of objects in the observable world. The kinetic theory of gases, mentioned above, is one example, and there are many others. Furthermore, such theories often have important technological applications. For example, laser technology is based on a theory about what happens when electrons in an atom go from higher to lower energy-states. And lasers work – they allow us to correct our vision, attack our enemies with guided missiles, and do much more besides. The theory that underpins laser technology is therefore highly empirically successful.

The empirical success of theories that posit unobservable entities is the basis of one of the strongest arguments for scientific realism, called the 'no miracles' argument. According to this argument, it would be an extraordinary coincidence if a theory that talks about electrons and atoms made accurate predictions about the observable world – unless electrons and atoms actually exist. If there are no atoms and electrons, what explains the theory's close fit with the observational data? Similarly, how do we explain the technological advances our theories have led to, unless by supposing that the theories in question are true? If atoms and electrons are just 'convenient fictions', as anti-realists maintain, then why do lasers work? On this view, being an anti-realist is akin to believing in miracles. Since it is obviously better not to believe in miracles if a non-miraculous alternative is available, we should be realists not anti-realists.

This argument is not intended to *prove* that realism is right and anti-realism wrong. Rather it is a plausibility argument – an inference to the best explanation. The phenomenon to be explained is the fact that many theories that postulate unobservable entities enjoy a high level of empirical success. The best explanation of this fact, say advocates of the 'no miracles' argument, is that the theories are true – the entities in question really exist, and behave just as the theories say. Unless we accept this explanation, the empirical success of our theories is an unexplained mystery.

Anti-realists have responded to the 'no miracles' argument in various ways. One response appeals to certain facts about the history of science. Historically, there are many cases of theories that we now believe to be false but that were empirically quite successful in their day. In a well-known article, the American philosopher of science Larry Laudan lists more than 30 such theories, drawn from a range of different scientific disciplines and eras. The phlogiston theory of combustion is one example. This theory, which was widely accepted until the end of the 18th century, held that when any object burns it releases a substance called 'phlogiston' into the atmosphere. Modern chemistry teaches us that this is false: there is no such substance as phlogiston. Rather, burning occurs when things react with oxygen in the air. But despite the non-existence of phlogiston, the phlogiston theory was empirically quite successful: it fitted the observational data available at the time reasonably well.

Examples of this sort suggest that the 'no miracles' argument for scientific realism is a bit too quick. Proponents of that argument regard the empirical success of today's scientific theories as evidence of their truth. But the history of science shows that empirically successful theories have often turned out to be false. So how do we know that the same fate will not befall today's theories? How do we know that the atomic theory of matter, for example, will not go the same way as the phlogiston theory? Once we pay due attention to the history of science, argue the anti-realists, we see that the inference from empirical success to theoretical truth is a very shaky one. The rational attitude towards the atomic theory is thus one of agnosticism – it may be true, or it may not. We just do not know, say the anti-realists.

This is a powerful counter to the 'no miracles' argument, but it is not completely decisive. Some realists have responded by modifying the argument slightly. According to the modified version, the empirical success of a theory is evidence that what the theory says about the unobservable world is approximately true, rather than precisely true. This weaker claim is less vulnerable to counter-examples from the history of science. It is also more modest: it allows the realist to admit that today's theories may not be correct down to every last detail, while still holding that they are broadly on the right lines. Another way of modifying the argument is by refining the notion of empirical success. Some realists hold that empirical success is not just a matter of fitting the known observational data, but rather allowing us to predict new observational phenomena that were previously unknown. Relative to this more stringent criterion of empirical success, it is less easy to find historical examples of empirically successful theories that later turned out to be false.

Whether these refinements can really save the 'no miracles' argument is debatable. They certainly reduce the number of historical counter-examples, but not to zero. One that remains is the wave theory of light, first put forward by Christian Huygens in 1690. According to this theory, light consists of wave-like vibrations in an invisible medium called the ether, which was supposed to permeate the whole universe. (The rival to the wave theory was the particle theory of light, favoured by Newton, which held that light consists of very small particles emitted by the light source.) The wave theory was not widely accepted until the French physicist Auguste Fresnel formulated a mathematical version of the theory in 1815, and used it to predict some surprising new optical phenomena. Optical experiments confirmed Fresnel's predictions, convincing many 19th-century scientists that the wave theory of light must be true. But modern physics tells us the theory is not true: there is no such thing as the ether, so light doesn't consist of vibrations in it. Again, we have an example of a false but empirically successful theory.

The important feature of this example is that it tells against even the modified version of the 'no miracles' argument. For Fresnel's theory *did* make novel predictions, so qualifies as empirically successful even relative to the stricter notion of empirical success. And it is hard to see how Fresnel's theory can be called 'approximately true', given that it was based around the idea of the ether, which does not exist. Whatever exactly it means for a theory to be approximately true, a necessary condition is surely that the entities the theory talks about really do exist. In short, Fresnel's theory was empirically successful even according to a strict understanding of this notion, but was not even approximately true. The moral of the story, say anti-realists, is that we should not assume that modern scientific theories are even roughly on the right lines, just because they are so empirically successful.

Whether the 'no miracles' argument is a good argument for scientific realism is therefore an open question. On the one hand, the argument is open to quite serious objections, as we have seen. On the other hand, there is something intuitively compelling about the argument. It really is hard to accept that atoms and electrons might not exist, when one considers the amazing success of theories that postulate these entities. But as the history of science shows, we should be very cautious about assuming that our current scientific theories are true, however well they fit the data. Many people have assumed that in the past and been proved wrong.

The observable/unobservable distinction

Central to the debate between realism and anti-realism is the distinction between things that are observable and things that are not. So far we have simply taken this distinction for granted – tables and chairs are observable, atoms and electrons are not. But in fact the distinction is quite philosophically problematic. Indeed, one of the main arguments for scientific realism says that it is not possible to draw the observable/unobservable distinction in a principled way.

Why should this be an argument for scientific realism? Because the coherence of anti-realism is crucially dependent on there being a clear distinction between the observable and the unobservable. Recall that anti-realists advocate a different attitude towards scientific claims, depending on whether they are about observable or unobservable parts of reality – we should remain agnostic about the truth of the latter, but not the former. Anti-realism thus presupposes that we can divide scientific claims into two sorts: those that are about observable entities and processes, and those that are not. If it turns out that this division cannot be made in a satisfactory way, then anti-realism is obviously in serious trouble, and realism wins by default. That is why scientific realists are often keen to emphasize the problems associated with the observable/unobservable distinction.

One such problem concerns the relation between observation and detection. Entities such as electrons are obviously not observable in the ordinary sense, but their presence can be detected using special pieces of apparatus called particle detectors. The simplest particle detector is the cloud chamber, which consists of a closed container filled with air that has been saturated with water-vapour (Figure 9). When charged particles such as electrons pass through the chamber, they collide with neutral atoms in the air, converting them into ions; water vapour condenses around these ions causing liquid droplets to form, which can be seen with the naked eye. We can follow the path of an electron through the cloud chamber by watching the tracks of these liquid droplets. Does this mean that electrons can be observed after all? Most philosophers would say no: cloud chambers allow us to detect electrons, not observe them directly. In much the same way, high-speed jets can be detected by the vapour trails they leave behind, but watching these trails is not observing the jet. But is it always clear how to distinguish observing from detecting? If not, then the anti-realist position could be in trouble.

In a well-known defence of scientific realism from the early 1960s, the American philosopher Grover Maxwell posed the following problem for the anti-realist. Consider the following sequence of events: looking at something with the naked eye, looking at something through a window, looking at something through a pair of strong glasses, looking at something through binoculars, looking at something though a low-powered microscope, looking at something through a high-powered microscope, and so on. Maxwell argued that these events lie on a smooth continuum. So how do we decide which count as observing and which not? Can a biologist observe micro-organisms with his high-powered microscope, or can he only detect their presence in the way that a physicist can detect the presence of electrons in a cloud chamber? If something can only be seen with the help of sophisticated scientific instruments, does it count as observable or unobservable? How sophisticated can the instrumentation be, before we have a case of detecting rather

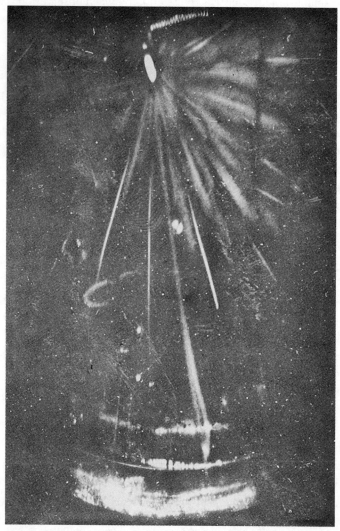

9. One of the first photographs to show the tracks of subatomic particles in a cloud chamber. The picture was taken by the cloud chamber's inventor, English physicist C. T. R. Wilson, at the Cavendish Laboratory in Cambridge in 1911. The tracks are due to alpha particles emitted by a small amount of radium on the top of a metal tongue inserted into the cloud chamber. As an electrically charged particle moves through the water vapour in a cloud chamber, it ionizes the gas, and water drops condense on the ions, thus producing a track of droplets where the particle has passed.

than observing? There is no principled way of answering these questions, Maxwell argued, so the anti-realist's attempt to classify entities as either observable or unobservable is doomed to failure.

Maxwell's argument is bolstered by the fact that scientists themselves sometimes talk about 'observing' particles with the help of sophisticated bits of apparatus. In the philosophical literature, electrons are usually taken as paradigm examples of unobservable entities, but scientists are often perfectly happy to talk about 'observing' electrons using particle detectors. Of course, this does not prove that the philosophers are wrong and that electrons are observable after all, for the scientists' talk is probably best regarded as a *façon-de-parler*. Similarly, the fact that scientists talk about having 'experimental proof' of a theory does not mean that experiments can really prove theories to be true, as we saw in Chapter 2. Nonetheless, if there really is a philosophically important observable/unobservable distinction, as anti-realists maintain, it is odd that it corresponds so badly with the way scientists themselves speak.

Maxwell's arguments are powerful, but by no means completely decisive. Bas van Fraassen, a leading contemporary anti-realist, claims that Maxwell's arguments only show 'observable' to be a vague concept. A vague concept is one that has borderline cases – cases that neither clearly do nor clearly do not fall under it. 'Bald' is an obvious example. Since hair loss comes in degrees, there are many men of whom it's hard to say whether they are bald or not. But van Fraassen points out that vague concepts are perfectly usable, and can mark genuine distinctions in the world. (In fact, most concepts are vague to at least some extent.) No-one would argue that the distinction between bald and hirsute men is unreal or unimportant simply because 'bald' is vague. Certainly, if we attempt to draw a sharp dividing line between bald and hirsute men, it will arbitrary. But since there are clear-cut cases of men who are bald and clear-cut cases of men who are not, the impossibility of drawing a sharp dividing line doesn't matter. The concept is perfectly usable despite its vagueness.

Precisely the same applies to 'observable', according to van Fraassen. There are clear-cut cases of entities that can be observed, for example chairs, and clear-cut cases of entities that cannot, for example electrons. Maxwell's argument highlights the fact that there are also borderline cases, where we are unsure whether the entities in question can be observed or only detected. So if we try to draw a sharp dividing line between observable and unobservable entities, it will inevitably be somewhat arbitrary. But as with baldness, this does not show that the observable/unobservable distinction is somehow unreal or unimportant, for there are clear-cut cases on either side. So the vagueness of the term 'observable' is no embarrassment to the anti-realist, van Fraassen argues. It only sets an upper limit on the precision with which she can formulate her position.

How strong an argument is this? Van Fraassen is certainly right that the existence of borderline cases, and the consequent impossibility of drawing a sharp boundary without arbitrariness, does not show the observable/unobservable distinction to be unreal. To that extent, his argument against Maxwell succeeds. However, it is one thing to show that there is a real distinction between observable and unobservable entities, and another to show that the distinction is capable of bearing the philosophical weight that anti-realists wish to place on it. Recall that anti-realists advocate an attitude of complete agnosticism towards claims about the unobservable part of reality – we have no way of knowing whether they are true or not, they say. Even if we grant van Fraassen his point that there are clear cases of unobservable entities, and that that is enough for the anti-realist to be getting on with, the anti-realist still needs to provide an argument for thinking that knowledge of unobservable reality is impossible.

The underdetermination argument

One argument for anti-realism centres on the relationship between scientists' observational data and their theoretical claims. Anti-realists emphasize that the ultimate data to which scientific theories are responsible is always observational in character. (Many realists would agree with this claim.) To illustrate, consider again the kinetic theory of gases, which says that any sample of gas consists of molecules in motion. Since these molecules are unobservable, we obviously cannot test the theory by directly observing various samples of gas. Rather, we need to deduce from the theory some statement that can be directly tested, which will invariably be about observable entities. As we saw, the kinetic theory implies that a sample of gas will expand when heated, if the pressure remains constant. This statement can be directly tested, by observing the readings on the relevant pieces of apparatus in a laboratory (Figure 10). This example illustrates a general truth: observational data

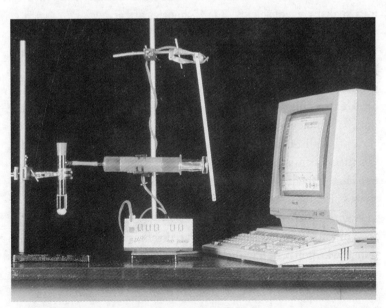

10. **Dialatometer for measuring the change in volume of a gas as its temperature varies.**

constitute the ultimate evidence for claims about unobservable entities.

Anti-realists then argue that the observational data 'underdetermine' the theories scientists put forward on their basis. What does this mean? It means that the data can in principle be explained by many different, mutually incompatible, theories. In the case of the kinetic theory, anti-realists will say that *one* possible explanation of the observational data is that gases contain large numbers of molecules in motion, as the kinetic theory says. But they will insist that there are other possible explanations too, which conflict with the kinetic theory. So according to anti-realists, scientific theories that posit unobservable entities are underdetermined by the observational data – there will always be a number of competing theories that can account for that data equally well.

It is easy to see why the underdetermination argument supports an anti-realist view of science. For if theories are always underdetermined by the observational data, how can we ever have reason to believe that a particular theory is true? Suppose a scientist advocates a given theory about unobservable entities, on the grounds that it can explain a large range of observational data. An anti-realist philosopher of science comes along, and argues that the data can in fact be accounted for by various alternative theories. If the anti-realist is correct, it follows that the scientist's confidence in her theory is misplaced. For what reason does the scientist have to choose the theory she does, rather than one of the alternatives? In such a situation, surely the scientist should admit that she has no idea which theory is true? Underdetermination leads naturally to the anti-realist conclusion that agnosticism is the correct attitude to take towards claims about the unobservable region of reality.

But is it actually true that a given set of observational data can always be explained by many different theories, as anti-realists maintain? Realists usually respond to the underdetermination argument by insisting that this claim is true only in a trivial and uninteresting sense. In principle, there will always be more than one possible explanation of a given set of observations. But, say the realists, it does not follow that all of these possible explanations are as good as one another. Just because two theories can both account for our observational data does not mean that there is nothing to choose between them. For one of the theories might be simpler than the other, for example, or might explain the data in a more intuitively plausible way, or might postulate fewer hidden causes, and so on. Once we acknowledge that there are criteria for theory choice in addition to compatibility with the observational data, the problem of underdetermination disappears. Not all the possible explanations of our observational data are as good as one another. Even if the data that the kinetic theory explains can in principle be explained by alternative theories, it does not follow that these alternatives can explain as well as the kinetic theory does.

This response to the underdetermination argument is bolstered by the fact that there are relatively few real cases of underdetermination in the history of science. If the observational data can always be explained equally well by many different theories, as anti-realists maintain, surely we should expect to find scientists in near perpetual disagreement with one another? But that is not what we find. Indeed, when we inspect the historical record, the situation is almost exactly the reverse of what the underdetermination argument would lead us to expect. Far from scientists being faced with a large number of alternative explanations of their observational data, they often have difficulty finding even *one* theory that fits the data adequately. This lends support to the realist view that underdetermination is merely a philosopher's worry, with little relation to actual scientific practice.

Anti-realists are unlikely to be impressed by this response. After all, philosophical worries are still genuine ones, even if their practical implications are few. Philosophy may not change the world, but that doesn't mean it isn't important. And the suggestion that criteria such as simplicity can be used to adjudicate between competing theories immediately invites the awkward question of why simpler theories should be thought more likely to be true; we touched on this issue in Chapter 2. Anti-realists typically grant that the problem of underdetermination can be eliminated in practice by using criteria such as simplicity to discriminate between competing explanations of our observational data. But they deny that such criteria are reliable indicators of the truth. Simpler theories may be more convenient to work with, but they are not intrinsically more probable than complex ones. So the underdetermination argument stands: there are always multiple explanations of our data, we have no way of knowing which is true, so knowledge of unobservable reality cannot be had.

However, the story does not end here; there is a further realist comeback. Realists accuse anti-realists of applying the underdetermination argument selectively. If the argument is applied consistently, it rules out not only knowledge of the unobservable world, but also knowledge of much of the observable world, say the realists. To understand why realists say this, notice that many things that are observable never actually get observed. For example, the vast majority of living organisms on the planet never get observed by humans, but they are clearly observable. Or think of an event such as a large meteorite hitting the earth. No-one has ever witnessed such an event, but it is clearly observable. It just so happens that no human was ever in the right place at the right time. Only a small fraction of what is observable actually gets observed.

The key point is this. Anti-realists claim that the unobservable part of reality lies beyond the limits of scientific knowledge. So they allow that we can have knowledge of objects and events that are observable but unobserved. But theories about unobserved objects and events are just as underdetermined by our data as are theories about unobservable ones. For example, suppose a scientist puts forward the hypothesis that a meteorite struck the moon in 1987.

He cites various pieces of observational data to support this hypothesis, e.g. that satellite pictures of the moon show a large crater that wasn't there before 1987. However, this data can in principle be explained by many alternative hypotheses – perhaps a volcanic eruption caused the crater, or an earthquake. Or perhaps the camera that took the satellite pictures was faulty, and there is no crater at all. So the scientist's hypothesis is underdetermined by the data, even though the hypothesis is about a perfectly observable event – a meteorite striking the moon. If we apply the underdetermination argument consistently, say realists, we are forced to conclude that we can only acquire knowledge of things that have actually been observed.

This conclusion is very implausible, and is not one that any philosopher of science would wish to accept. For much of what scientists tell us concerns things that have not been observed – think of ice ages, dinosaurs, continental drift, and the like. To say that knowledge of the unobserved is impossible is to say that most of what passes for scientific knowledge is not really knowledge at all. Of course, scientific realists do not accept this conclusion. Rather, they take it as evidence that the underdetermination argument must be wrong. Since science clearly does give us knowledge of the unobserved, despite the fact that theories about the unobserved are underdetermined by our data, it follows that underdetermination is no barrier to knowledge. So the fact that our theories about the unobservable are also underdetermined by our data does not mean that science cannot give us knowledge of the unobservable region of the world.

In effect, realists who argue this way are saying that the problem raised by the underdetermination argument is simply a sophisticated version of the problem of induction. To say that a theory is underdetermined by the data is to say that there are alternative theories that can account for the same data. But this is effectively just to say that the data do not entail the theory: the inference from the data to the theory is non-deductive. Whether the theory is about unobservable entities, or about observable but unobserved entities, makes no difference – the logic of the situation is the same in both cases. Of course, showing that the underdetermination argument is just a version of the problem of induction does not mean that it can be ignored. For there is little consensus on how the problem of induction should be tackled, as we saw in Chapter 2. But it does mean that there is no *special* difficulty about unobservable entities. Therefore the anti-realist position is ultimately arbitrary, say the realists. Whatever problems there are in understanding how science can give us knowledge of atoms and electrons are equally problems for understanding how science can give us knowledge of ordinary, medium-sized objects.

Chapter 5
Scientific change and scientific revolutions

Scientific ideas change fast. Pick virtually any scientific discipline you like, and you can be sure that the prevalent theories in that discipline will be very different from those of 50 years ago, and extremely different from those of 100 years ago. Compared with other areas of intellectual endeavour such as philosophy and the arts, science is a rapidly changing activity. A number of interesting philosophical questions centre on the issue of scientific change. Is there a discernible pattern to the way scientific ideas change over time? When scientists abandon their existing theory in favour of a new one, how should we explain this? Are later scientific theories objectively better than earlier ones? Or does the concept of objectivity make sense at all?

Most modern discussion of these questions takes off from the work of the late Thomas Kuhn, an American historian and philosopher of science. In 1963 Kuhn published a book called *The Structure of Scientific Revolutions*, unquestionably the most influential work of philosophy of science in the last 50 years. The impact of Kuhn's ideas has also been felt in other academic disciplines such as sociology and anthropology, and in the general intellectual culture at large. (*The Guardian* newspaper included *The Structure of Scientific Revolutions* in its list of the 100 most influential books of the 20th century.) In order to understand why Kuhn's ideas caused such a stir, we need to look briefly at the state of philosophy of science prior to the publication of his book.

Logical positivist philosophy of science

The dominant philosophical movement in the English-speaking world in the post-war period was *logical positivism*. The original logical positivists were a loosely knit group of philosophers and scientists who met in Vienna in the 1920s and early 1930s, under the leadership of Moritz Schlick. (Carl Hempel, whom we met in Chapter 3, was closely associated with the positivists, as was Karl Popper.) Fleeing persecution by the Nazis, most of the positivists emigrated to the United States, where they and their followers exerted a powerful influence on academic philosophy until about the mid-1960s, by which time the movement had begun to disintegrate.

The logical positivists had a very high regard for the natural sciences, and also for mathematics and logic. The early years of the 20th century witnessed exciting scientific advances, particularly in physics, which impressed the positivists tremendously. One of their aims was to make philosophy itself more 'scientific', in the hope that this would allow similar advances to be made in philosophy. What particularly impressed the positivists about science was its apparent objectivity. Unlike in other fields, where much turned on the subjective opinion of enquirers, scientific questions could be settled in a fully objective way, they believed. Techniques such as experimental testing allowed a scientist to compare his theory directly with the facts, and thus reach an informed, unbiased decision about the theory's merits. Science for the positivists was thus a paradigmatically rational activity, the surest route to the truth that there is.

Despite the high esteem in which they held science, the positivists paid little attention to the history of science. Indeed, they believed that philosophers had little to learn from studying history of science. This was primarily because they drew a sharp distinction between what they called the 'context of discovery' and the 'context of justification'. The context of discovery refers to the actual historical process by which a scientist arrives at a given theory. The context of justification refers to the means by which the scientist tries to justify his theory once it is already there – which includes testing the theory, searching for relevant evidence, and so on. The positivists believed that the former was a subjective, psychological process that wasn't governed by precise rules, while the latter was an objective matter of logic. Philosophers of science should confine themselves to studying the latter, they argued.

An example can help make this idea clearer. In 1865 the Belgian scientist Kekule discovered that the benzene molecule has a hexagonal structure. Apparently, he hit on the hypothesis of a hexagonal structure for benzene after a dream in which he saw a snake trying to bite its own tail (Figure 11). Of course, Kekule then had to test his hypothesis scientifically, which he did. This is an extreme example, but it shows that scientific hypotheses can be arrived at in the most unlikely of ways – they are not always the product of careful, systematic thought. The positivists would argue that it makes no difference how a hypothesis is arrived at initially. What matters is how it is tested once it is already there – for it is this that makes science a rational activity. How Kekule first arrived at his hypothesis was immaterial; what mattered was how he justified it.

This sharp distinction between discovery and justification, and the belief that the former is 'subjective' and 'psychological' while the latter is not, explains why the positivists' approach to philosophy of science was so ahistorical. For the actual historical process by which scientific ideas change and develop lies squarely in the context of discovery, not the context of justification. That process might be of interest to historians or psychologists, but had nothing to teach philosophers of science, according to the positivists.

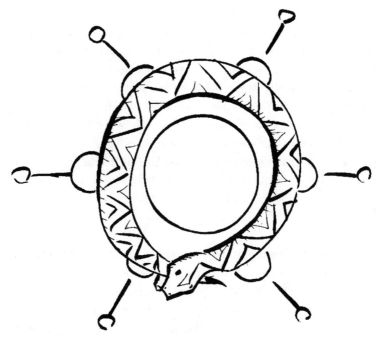

11. Kekule arrived at the hypothesis of the hexagonal structure of benzene after a dream in which he saw a snake trying to bite its own tail.

Another important theme in positivist philosophy of science was the distinction between theories and observational facts; this is related to the observable/unobservable distinction discussed in the previous chapter. The positivists believed that disputes between rival scientific theories could be solved in a perfectly objective way – by comparing the theories directly with the 'neutral' observational facts, which all parties could accept. The positivists disagreed between themselves about how exactly this set of neutral facts should be characterized, but they were adamant that it existed. Without a clear distinction between theories and observational facts, the rationality and objectivity of science would be compromised, and the positivists were resolute in their belief that science was rational and objective.

The structure of scientific revolutions

Kuhn was a historian of science by training, and firmly believed that philosophers had much to learn from studying the history of science. Insufficient attention to the history of science had led the positivists to form an inaccurate and naïve picture of the scientific enterprise, he maintained. As the title of his book indicates, Kuhn was especially interested in scientific revolutions – periods of great upheaval when existing scientific ideas are replaced with radically new ones. Examples of scientific revolutions are the Copernican revolution in astronomy, the Einsteinian revolution in physics, and the Darwinian revolution in biology. Each of these revolutions led to a fundamental change in the scientific world-view – the overthrow of an existing set of ideas by a completely different set.

Of course, scientific revolutions happen relatively infrequently – most of the time any given science is not in a state of revolution. Kuhn coined the term 'normal science' to describe the ordinary day-to-day activities that scientists engage in when their discipline is not undergoing revolutionary change. Central to Kuhn's account of normal science is the concept of a *paradigm*. A paradigm consists of two main components: firstly, a set of fundamental theoretical assumptions that all members of a scientific community accept at a given time; secondly, a set of 'exemplars' or particular scientific problems that have been solved by means of those theoretical assumptions, and that appear in the textbooks of the discipline in question. But a paradigm is more than just a theory (though Kuhn sometimes uses the words interchangeably). When scientists share a paradigm they do not just agree on certain scientific propositions, they agree also on how future scientific research in their field should proceed, on which problems are the pertinent ones to tackle, on what the appropriate methods for solving those problems are, on what an acceptable solution of the problems would look like, and so on. In short, a paradigm is an entire scientific outlook – a constellation of shared assumptions, beliefs, and values that unite a scientific community and allow normal science to take place.

What exactly does normal science involve? According to Kuhn it is primarily a matter of *puzzle-solving*. However successful a paradigm is, it will always encounter certain problems – phenomena that it cannot easily accommodate, mismatches between the theory's predictions and the experimental facts, and so on. The job of the normal scientist is to try to eliminate these minor puzzles while making as few changes as possible to the paradigm. So normal science is a highly conservative activity – its practitioners are not trying to make any earth-shattering discoveries, but rather just to develop and extend the existing paradigm. In Kuhn's words, 'normal science does not aim at novelties of fact or theory, and when successful finds none'. Above all, Kuhn stressed that normal scientists are not trying to *test* the paradigm. On the contrary, they accept the paradigm unquestioningly, and conduct their research within the limits it sets. If a normal scientist gets an experimental result that conflicts with the paradigm, she will usually assume that her experimental technique is faulty, not that the paradigm is wrong. The paradigm itself is not negotiable.

Typically, a period of normal science lasts many decades, sometimes even centuries. During this time scientists gradually articulate the paradigm – fine-tuning it, filling in details, solving more and more puzzles, extending its range of application, and so on. But over time *anomalies* are discovered – phenomena that simply cannot be reconciled with the theoretical assumptions of the paradigm, however hard normal scientists try. When anomalies are few in number they tend to just get ignored. But as more and more anomalies accumulate, a burgeoning sense of crisis envelops the scientific community. Confidence in the existing paradigm breaks down, and the process of normal science temporarily grinds to a halt. This marks the beginning of a period of 'revolutionary science' as Kuhn calls it. During such periods, fundamental scientific ideas *are* up for grabs. A variety of alternatives to the old paradigm are proposed, and eventually a new paradigm becomes established. A generation or so is usually required before all members of the scientific community are won over to the new paradigm – an event that marks the completion of a scientific revolution. The essence of a scientific revolution is thus the shift from an old paradigm to a new one.

Kuhn's characterization of the history of science as long periods of normal science punctuated by occasional scientific revolutions struck a chord with many philosophers and historians of science. A number of examples from the history of science fit Kuhn's model quite well. When we examine the transition from Ptolemaic to Copernican astronomy, for example, or from Newtonian to Einsteinian physics, many of the features that Kuhn describes are present. Ptolemaic astronomers did indeed share a paradigm, based around the theory that the earth is stationary at the centre of the universe, which formed the unquestioned back-drop to their investigations. The same is true of Newtonian physicists in the 18th and 19th centuries, whose paradigm was based around Newton's theory of mechanics and gravitation. And in both cases, Kuhn's account of how an old paradigm gets replaced by a new one applies fairly accurately. There are also scientific revolutions that do not fit the Kuhnian model so neatly – for example the recent molecular revolution in biology. But nonetheless, most people agree that Kuhn's description of the history of science contains much of value.

Why did Kuhn's ideas cause such a storm? Because in addition to his purely descriptive claims about the history of science, Kuhn advanced some highly controversial philosophical theses. Ordinarily we assume that when scientists trade their existing theory for a new one, they do so on the basis of objective evidence. But Kuhn argued that adopting a new paradigm involves a certain act of faith on the part of the scientist. He allowed that a scientist could have good reasons for abandoning an old paradigm for a new one, but he insisted that reasons alone could never rationally *compel* a paradigm shift. 'The transfer of allegiance from paradigm to paradigm', Kuhn wrote, 'is a conversion experience which cannot be forced'. And in explaining why a new paradigm rapidly gains acceptance in the scientific community, Kuhn emphasized the peer pressure of scientists on one another. If a given paradigm has very forceful advocates, it is more likely to win widespread acceptance.

Many of Kuhn's critics were appalled by these claims. For if paradigm shifts work the way Kuhn says, it is hard to see how science can be regarded as a rational activity at all. Surely scientists are meant to base their beliefs on evidence and reason, not on faith and peer pressure? Faced with two competing paradigms, surely the scientist should make an objective comparison of them to determine which has more evidence in its favour? Undergoing a 'conversion experience', or allowing oneself to be persuaded by the most forceful of one's fellow scientists, hardly seems like a rational way to behave. Kuhn's account of paradigm shifts seems hard to reconcile with the familiar positivist image of science as an objective, rational activity. One critic wrote that on Kuhn's account, theory choice in science was 'a matter for mob psychology'.

Kuhn also made some controversial claims about the overall direction of scientific change. According to a widely held view, science progresses towards the truth in a linear fashion, as older incorrect ideas get replaced by newer, correct ones. Later theories are thus objectively better than earlier ones. This 'cumulative' conception of science is popular among laymen and scientists alike, but Kuhn argued that it is both historically inaccurate and philosophically naïve. For example, he noted that Einstein's theory of relativity is in some respects more similar to Aristotelian than Newtonian theory – so the history of mechanics is not simply a linear progression from wrong to right. Moreover, Kuhn questioned whether the concept of objective truth actually makes sense at all. The idea that there is a fixed set of facts about the world, independent of any particular paradigm, was of dubious coherence, he believed. Kuhn suggested a radical alternative: the facts about the world are paradigm-relative, and thus change when paradigms change. If this suggestion is right, then it makes no sense to ask whether a given theory corresponds to the facts 'as they really are', nor therefore to ask whether it is objectively true. Truth itself becomes relative to a paradigm.

Incommensurability and the theory-ladenness of data

Kuhn had two main philosophical arguments for these claims. Firstly, he argued that competing paradigms are typically 'incommensurable' with one another. To understand this idea, we must remember that for Kuhn a scientist's paradigm determines her entire world-view – she views everything through the paradigm's lens. So when an existing paradigm is replaced by a new one in a scientific revolution, scientists have to abandon the whole conceptual framework which they use to make sense of the world. Indeed, Kuhn even claims, obviously somewhat metaphorically, that before and after a paradigm shift scientists 'live in different worlds'. Incommensurability is the idea that two paradigms may be so different as to render impossible any straightforward comparison of them with each other – there is no common language into which both can be translated. As a result, the proponents of different paradigms 'fail to make complete contact with each other's viewpoints', Kuhn claimed.

This is an interesting if somewhat vague idea. The doctrine of incommensurability stems largely from Kuhn's belief that scientific concepts derive their meaning from the theory in which they play a role. So to understand Newton's concept of mass, for example, we need to understand the whole of Newtonian theory – concepts cannot be explained independently of the theories in which they are embedded. This idea, which is sometimes called 'holism', was taken very seriously by Kuhn. He argued that the term 'mass' actually meant something different for Newton and Einstein, since the theories in which each embedded the term were so different. This implies that Newton and Einstein were in effect speaking different languages, which obviously complicates the attempt to choose between their theories. If a Newtonian and an Einsteinian physicist tried to have a rational discussion, they would end up talking past each other.

Kuhn used the incommensurability thesis both to rebut the view that paradigm shifts are fully 'objective', and to bolster his non-cumulative picture of the history of science. Traditional philosophy of science saw no huge difficulty in choosing between competing theories – you simply make an objective comparison of them, in the light of the available evidence, and decide which is better. But this clearly presumes that there is a common language in which both theories can be expressed. If Kuhn is right that proponents of old and new paradigms are quite literally talking past each other, no such simplistic account of paradigm choice can be correct. Incommensurability is equally problematic for the traditional 'linear' picture of scientific history. If old and new paradigms are incommensurable, then it cannot be correct to think of scientific revolutions as the replacement of 'wrong' ideas by 'right' ones. For to call one idea right and another wrong implies the existence of a common framework for evaluating them, which is precisely what Kuhn denies. Incommensurability implies that scientific change, far from being a straightforward progression towards the truth, is in a sense directionless: later paradigms are not better than earlier ones, just different.

Not many philosophers were convinced by Kuhn's incommensurability thesis. Part of the problem was that Kuhn also claimed old and new paradigms to be *incompatible*. This claim is very plausible, for if old and new paradigms were not incompatible there would be no need to choose between them. And in many cases the incompatibility is obvious – the Ptolemaic claim that the planets revolve around the earth is obviously incompatible with the Copernican claim that they revolve around the sun. But as Kuhn's critics were quick to point out, if two things are incommensurable then they cannot be incompatible. To see why not, consider the proposition that an object's mass depends on its velocity. Einstein's theory says this proposition is true while Newton's says it is false.

But if the doctrine of incommensurability is right, then there is no actual disagreement between Newton and Einstein here, for the proposition means something different for each. Only if the proposition has the *same* meaning in both theories, i.e. only if there is no incommensurability, is there a genuine conflict between the two. Since everybody (including Kuhn) agrees that Einstein's and Newton's theories do conflict, that is strong reason to regard the incommensurability thesis with suspicion.

In response to objections of this type, Kuhn moderated his incommensurability thesis somewhat. He insisted that even if two paradigms were incommensurable, that did not mean it was impossible to compare them with each other; it only made comparison more difficult. *Partial* translation between different paradigms could be achieved, Kuhn argued, so the proponents of old and new paradigms could communicate to some extent: they would not always be talking past each other entirely. But Kuhn continued to maintain that fully objective choice between paradigms was impossible. For in addition to the incommensurability deriving from the lack of a common language, there is also what he called 'incommensurability of standards'. This is the idea that proponents of different paradigms may disagree about the standards for evaluating paradigms, about which problems a good paradigm should solve, about what an acceptable solution to those problems would look like, and so on. So even if they can communicate effectively, they will not be able to reach agreement about whose paradigm is superior. In Kuhn's words, 'each paradigm will be shown to satisfy the criteria that it dictates for itself and to fall short of a few of those dictated by its opponent'.

Kuhn's second philosophical argument was based on an idea known as the 'theory-ladenness' of data. To grasp this idea, suppose you are a scientist trying to choose between two conflicting theories. The obvious thing to do is to look for a piece of data that will decide between the two – which is just what traditional philosophy of science recommended. But this will only be possible if there exist data that are suitably independent of the theories, in the sense that a scientist would accept the data whichever of the two theories she believed. As we have seen, the logical positivists believed in the existence of such theory-neutral data, which could provide an objective court of appeal between competing theories. But Kuhn argued that the ideal of theory-neutrality is an illusion – data are invariably contaminated by theoretical assumptions. It is impossible to isolate a set of 'pure' data which all scientists would accept irrespective of their theoretical persuasion.

The theory-ladenness of data had two important consequences for Kuhn. Firstly, it meant that the issue between competing paradigms could not be resolved by simply appealing to 'the data' or 'the facts', for what a scientist counts as data, or facts, will depend on which paradigm she accepts. Perfectly objective choice between two paradigms is therefore impossible: there is no neutral vantage-point from which to assess the claims of each. Secondly, the very idea of objective truth is called into question. For to be objectively true, our theories or beliefs must correspond to the facts, but the idea of such a correspondence makes little sense if the facts themselves are infected by our theories. This is why Kuhn was led to the radical view that truth itself is relative to a paradigm.

Why did Kuhn think that all data are theory-laden? His writings are not totally clear on this point, but at least two lines of argument are discernible. The first is the idea that perception is heavily conditioned by background beliefs – what we see depends in part on what we believe. So a trained scientist looking at a sophisticated piece of apparatus in a laboratory will see something different from what a layman sees, for the scientist obviously has many beliefs about the apparatus that the layman lacks. There are a number of psychological experiments that supposedly show that perception is sensitive in this way to background belief – though the correct interpretation of these experiments is a contentious matter. Secondly, scientists' experimental and observational reports are often couched in highly theoretical language. For example, a scientist might report the outcome of an experiment by saying 'an electric current is flowing through the copper rod'. But this data report is obviously laden with a large amount of theory. It would not be accepted by a scientist who did not hold standard beliefs about electric currents, so it is clearly not theory-neutral.

Philosophers are divided over the merits of these arguments. On the one hand, many agree with Kuhn that pure theory-neutrality is an unattainable ideal. The positivists' idea of a class of data statements totally free of theoretical commitment is rejected by most contemporary philosophers – not least because no-one has succeeded in saying what such statements would look like. But it is not clear that this compromises the objectivity of paradigm shifts altogether. Suppose, for example, that a Ptolemaic and a Copernican astronomer are engaged in a debate about whose theory is superior. In order for them to debate meaningfully, there needs to be some astronomical data they can agree on. But why should this be a problem? Surely they can agree about the relative position of the earth and the moon on successive nights, for example, or the time at which the sun rises? Obviously, if the Copernican insists on describing the data in a way that presumes the truth of the heliocentric theory, the Ptolemaist will object. But there is no reason why the Copernican should do that. Statements such as 'on May 14th the sun rose at 7.10 a.m.' can be agreed on by a scientist whether they believe the geocentric or the heliocentric theory. Such statements may not be *totally* theory-neutral, but they are sufficiently free of theoretical contamination to be acceptable to proponents of both paradigms, which is what matters.

It is even less obvious that the theory-ladenness of data forces us to abandon the concept of objective truth. Many philosophers would accept that theory-ladenness makes it hard to see how *knowledge* of objective truth is possible, but that is not to say that the very concept is incoherent. Part of the problem is that, like many people who are suspicious of the concept of objective truth, Kuhn failed to articulate a viable alternative. The radical view that truth is paradigm-relative is ultimately hard to make sense of. For like all such relativist doctrines, it faces a critical problem. Consider the question: is the claim that truth is paradigm-relative *itself* objectively true or not? If the proponent of relativism answers 'yes', then they have admitted that the concept of objective truth does make sense and have thus contradicted themselves. If they answer 'no', then they have no grounds on which to argue with someone who disagrees and says that, in their opinion, truth is *not* paradigm-relative. Not all philosophers regard this argument as completely fatal to relativism, but it does suggest that abandoning the concept of objective truth is easier said than done. Kuhn certainly raised some telling objections to the traditional view that the history of science is simply a linear progression to the truth, but the relativist alternative he offered in its place is far from unproblematic.

Kuhn and the rationality of science

The Structure of Scientific Revolutions is written in a very radical tone. Kuhn gives every impression of wanting to replace standard philosophical ideas about theory change in science with a totally new conception. His doctrine of paradigm shifts, of incommensurability, and of the theory-ladenness of data seems wholly at odds with the positivist view of science as a rational, objective, and cumulative enterprise. With much justification, most of Kuhn's early readers took him to be saying that science is an entirely non-rational activity, one characterized by dogmatic adherence to a paradigm in normal periods, and sudden 'conversion experiences' in revolutionary periods.

But Kuhn himself was unhappy with this interpretation of his work. In a Postscript to the second edition of *The Structure of Scientific Revolutions* published in 1970, and in subsequent writings, Kuhn moderated his tone considerably – and accused some of his early readers of having misread his intentions. His book was not an attempt to cast doubt on the rationality of science, he argued, but rather to offer a more realistic, historically accurate picture of how science actually develops. By neglecting the history of science, the positivists had been led to an excessively simplistic, indeed idealistic, account of how science works, and Kuhn's aim was simply to provide a corrective. He was not trying to show that science was irrational, but rather to provide a better account of what scientific rationality involves.

Some commentators regard Kuhn's Postscript as simply an about-turn – a retreat from his original position, rather than a clarification of it. Whether this is a fair assessment is not a question we will go into here. But the Postscript did bring to light one important issue. In rebutting the charge that he had portrayed paradigm shifts as non-rational, Kuhn made the famous claim that there is 'no algorithm' for theory choice in science. What does this mean? An algorithm is of a set of rules that allows us to compute the answer to a particular question. For example, an algorithm for multiplication is a set of rules that when applied to any two numbers tells us their product. (When you learn arithmetic in primary school, you in effect learn algorithms for addition, subtraction, multiplication, and division.) So an algorithm for theory choice is a set of rules that when applied to two competing theories would tell us which we should choose. Much positivist philosophy of science was in effect committed to the existence of such an algorithm. The positivists often wrote as if, given a set of data and two competing theories, the 'principles of scientific method' could be used to determine which theory was superior. This idea was implicit in their belief that although discovery was a matter of psychology, justification was a matter of logic.

Kuhn's insistence that there is no algorithm for theory choice in science is almost certainly correct. For no-one has ever succeeded in producing such an algorithm. Lots of philosophers and scientists have made plausible suggestions about what to look for in theories – simplicity, broadness of scope, close fit with the data, and so on. But these suggestions fall far short of providing a true algorithm, as Kuhn knew well. For one thing, there may be trade-offs: theory one may be simpler than theory two, but theory two may fit the data more closely. So an element of subjective judgement, or scientific common-sense, will often be needed to decide between competing theories. Seen in this light Kuhn's suggestion that the adoption of a new paradigm involves a certain act of faith does not seem quite so radical, and likewise his emphasis on the persuasiveness of a paradigm's advocates in determining its chance of winning over the scientific community.

The thesis that there is no algorithm for theory choice lends support to the view that Kuhn's account of paradigm shifts is not an assault on the rationality of science. For we can read Kuhn instead as rejecting a certain conception of rationality. The positivists believed, in effect, that there *must* be an algorithm for theory choice on pain of scientific change being irrational. This is by no means a crazy view: many paradigm cases of rational action do involve rules, or algorithms. For example, if you want to decide whether a good is cheaper in England or Japan, you apply an algorithm for converting pounds into yen; any other way of trying to decide the matter is irrational. Similarly, if a scientist is trying to decide between two competing theories, it is tempting to think that the only rational way to proceed is to apply an algorithm for theory choice. So if it turns out that there is no such algorithm, as seems likely, we have two options. Either we can conclude that scientific change is irrational *or* that the positivist conception of rationality is too demanding. In the Postscript Kuhn suggests that the latter is the correct reading of his work. The moral of his story is not that paradigm shifts are irrational, but rather that a more relaxed, non-algorithmic concept of rationality is required to make sense of them.

Kuhn's legacy

Despite their controversial nature, Kuhn's ideas transformed philosophy of science. In part this is because Kuhn called into question many assumptions that had traditionally been taken for granted, forcing philosophers to confront them, and in part because he drew attention to a range of issues that traditional philosophy of science had simply ignored. After Kuhn, the idea that philosophers could afford to ignore the history of science appeared increasingly untenable, as did the idea of a sharp dichotomy between the contexts of discovery and justification. Contemporary philosophers of science pay much greater attention to the historical development of science than did their pre-Kuhnian ancestors. Even those unsympathetic to Kuhn's more radical ideas would accept that in these respects his influence has been positive.

Another important impact of Kuhn's work was to focus attention on the social context in which science takes place, something that traditional philosophy of science ignored. Science for Kuhn is an intrinsically social activity: the existence of a scientific community, bound together by allegiance to a shared paradigm, is a pre-requisite for the practice of normal science. Kuhn also paid considerable attention to how science is taught in schools and universities, how young scientists are initiated into the scientific community, how scientific results are published, and other such 'sociological' matters. Not surprisingly, Kuhn's ideas have been very influential among sociologists of science. In particular, a movement known as the 'strong programme' in the sociology of science, which emerged in Britain in the 1970s, owed much to Kuhn.

The strong programme was based around the idea that science should be viewed as a product of the society in which it is practised. Strong programme sociologists took this idea very literally: they held that scientists' beliefs were in large part socially determined. So to explain why a scientist believes a given theory, for example, they would cite aspects of the scientist's social and cultural background. The scientist's own reasons for believing the theory were never explanation enough, they maintained. The strong programme borrowed a number of themes from Kuhn, including the theory-ladenness of data, the view of science as an essentially social enterprise, and the idea that there is no algorithm for theory choice. But strong programme sociologists were more radical than Kuhn, and less cautious. They openly rejected the notions of objective truth and rationality, which they regarded as ideologically suspect, and viewed traditional philosophy of science with great suspicion. This led to a certain amount of tension between philosophers and sociologists of science, which continues to this day.

Further afield, Kuhn's work has played a role in the rise of *cultural relativism* in the humanities and social sciences. Cultural relativism is not a precisely defined doctrine, but the central idea is that there is no such thing as absolute truth – truth is always relative to a particular culture. We may think that Western science reveals the truth about the world, but cultural relativists would say that other cultures and societies, for example indigenous Americans, have their own truth. As we have seen, Kuhn did indeed embrace relativist ideas. However, there is actually a certain irony in his having influenced cultural relativism. For cultural relativists are normally very anti-science. They object to the exalted status that science is accorded in our society, arguing that it discriminates against alternative belief systems that are equally valuable. But Kuhn himself was strongly pro-science. Like the positivists, he regarded modern science as a hugely impressive intellectual achievement. His doctrine of paradigm shifts, of normal and revolutionary science, of incommensurability and of theory-ladenness was not intended to undermine or criticize the scientific enterprise, but rather to help us understand it better.

Chapter 6
Philosophical problems in physics, biology, and psychology

The issues we have studied so far – induction, explanation, realism, and scientific change – belong to what is called 'general philosophy of science'. These issues concern the nature of scientific investigation in general, rather than pertaining specifically to chemistry, say, or geology. However, there are also many interesting philosophical questions that are specific to particular sciences – they belong to what is called 'philosophy of the special sciences'. These questions usually depend partly on philosophical considerations and partly on empirical facts, which is what makes them so interesting. In this chapter we examine three such questions, one each from physics, biology, and psychology.

Leibniz versus Newton on absolute space

Our first topic is a debate between Gottfried Leibniz (1646–1716) and Isaac Newton (1642–1727), two of the outstanding scientific intellects of the 17th century, concerning the nature of space and time. We shall focus primarily on space, but the issues about time are closely parallel. In his famous *Principles of Natural Philosophy*, Newton defended what is called an 'absolutist' conception of space. According to this view, space has an 'absolute' existence over and above the spatial relations between objects. Newton thought of space as a three-dimensional container into which God had placed the material universe at creation. This implies that space existed before there were any material objects, just as a container like a cereal box exists before any pieces of cereal are put inside. The only difference between space and ordinary containers like cereal boxes, according to Newton, is that the latter obviously have finite dimensions, whereas space extends infinitely in every direction.

Leibniz strongly disagreed with the absolutist view of space, and with much else in Newton's philosophy. He argued that space consists simply of the totality of spatial relations between material objects. Examples of spatial relations are 'above', 'below', 'to the left of', and 'to the right of' – they are relations that material objects bear to each other. This 'relationist' conception of space implies that before there were any material objects, space did not exist. Leibniz argued that space came into existence *when* God created the material universe; it did not exist beforehand, waiting to be filled up with material objects. So space is not usefully thought of as a container, nor indeed as an entity of any sort. Leibniz's view can be understood in terms of an analogy. A legal contract consists of a relationship between two parties – the buyer and seller of a house, for example. If one of the parties dies, then the contract ceases to exist. So it would be crazy to say that the contract has an existence independently of the relationship between buyer and seller – the contract just *is* that relationship. Similarly, space is nothing over and above the spatial relations between objects.

Newton's main reason for introducing the concept of absolute space was to distinguish between absolute and relative motion. Relative motion is the motion of one object with respect to another. So far as relative motion is concerned, it makes no sense to ask whether an object is 'really' moving or not – we can only ask whether it is moving with respect to some other object. To illustrate, imagine two joggers running in tandem along a straight road. Relative to a by-stander standing on the roadside, both are obviously in motion:

they are getting further away by the moment. But relative to each other, the joggers are not in motion: their relative positions remain exactly the same, so long as they keep jogging in the same direction at the same speed. So an object may be in relative motion with respect to one thing but be stationary with respect to another.

Newton believed that as well as relative motion, there is also absolute motion. Common-sense supports this view. For intuitively, it *does* make sense to ask whether an object is 'really' moving or not. Imagine two objects in relative motion – say a hang-glider and an observer on the earth. Now relative motion is symmetric: just as the hang-glider is in motion relative to the observer on the earth, so the observer is in motion relative to the hang-glider. But surely it makes sense to ask whether the observer or the hang-glider is 'really' moving, or both? If that is so, then we need the concept of absolute motion.

But what exactly *is* absolute motion? According to Newton, it is the motion of an object *with respect to absolute space itself.* Newton thought that at any time, every object has a particular location in absolute space. If an object changes its location in absolute space from one time to another then it is in absolute motion; otherwise, it is at absolute rest. So we need to think of space as an absolute entity, over and above the relations between material objects, in order to distinguish relative from absolute motion. Notice that Newton's reasoning rests on an important assumption. He assumes without question that all motion has got to be relative to something. Relative motion is motion relative to other material objects; absolute motion is motion relative to absolute space itself. So in a sense, even absolute motion is 'relative' for Newton. In effect, Newton is assuming that being in motion, whether absolute or relative, cannot be a 'brute fact' about an object; it can only be a fact about the object's relations to something else. That something else can either be another material object, or it can be absolute space.

Leibniz accepted that there was a difference between relative and absolute motion, but he denied that the latter should be explained as motion with respect to absolute space. For he regarded the concept of absolute space as incoherent. He had a number of arguments for this view, many of which were theological in nature. From a philosophical point of view, Leibniz's most interesting argument was that absolute space conflicts with what he called the principle of the identity of indiscernibles (PII). Since Leibniz regarded this principle as indubitably true, he rejected the concept of absolute space.

PII says that if two objects are indiscernible, then they are identical, i.e. they are really one and the same object. What does it mean to call two objects indiscernible? It means that no difference at all can be found between them – they have exactly the same attributes. So if PII is true, then any two genuinely distinct objects must differ in at least one of their attributes – otherwise they would be one, not two. PII is intuitively quite compelling. It certainly is not easy to find an example of two distinct objects that share *all* their attributes. Even two mass-produced factory goods will normally differ in innumerable ways, even if the differences cannot be detected with the naked eye. Whether PII is true in general is a complex question that philosophers still debate; the answer depends in part on exactly what counts as an 'attribute', and in part on difficult issues in quantum physics. But our concern for the moment is the use to which Leibniz puts the principle.

Leibniz uses two thought experiments to reveal a conflict between Newton's theory of absolute space and PII. His argumentative strategy is indirect: he assumes for the sake of argument that Newton's theory is correct, then tries to show that a contradiction follows from that assumption; since contradictions cannot be true, Leibniz concludes that Newton's theory must be false. Recall that for Newton, at any moment in time every object in the universe has a definite location in absolute space. Leibniz asks us to imagine two different universes, both containing exactly the same objects. In universe one, each object occupies a particular location in absolute space. In universe two, each object has been shifted to a different location in absolute space, two miles to the east (for example). There would be no way of telling these two universes apart. For we cannot observe the position of an object in absolute space, as Newton himself admitted. All we can observe are the positions of objects *relative to each other*, and these would remain unchanged – for all objects are shifted by the same amount. No observations or experiments could ever reveal whether we lived in universe one or two.

The second thought experiment is similar. Recall that for Newton, some objects are moving through absolute space while others are at rest. This means that at each moment, every object has a definite absolute velocity. (Velocity is speed in a given direction, so an object's absolute velocity is the speed at which it moves through absolute space in a specified direction. Objects at absolute rest have an absolute velocity of zero.) Now imagine two different universes, both containing exactly the same objects. In universe one, each object has a particular absolute velocity. In universe two, the absolute velocity of each object has been boosted by a fixed amount, say 300 kilometres per hour in a specified direction. Again, we could never tell these two universes apart. For it is impossible to observe how fast an object is moving with respect to absolute space, as Newton himself admitted. We can only observe how fast objects are moving *relative to each other* – and these relative velocities would remain unchanged, for the velocity of every object is boosted by exactly the same amount. No observations or experiments could ever reveal whether we lived in universe one or two.

In each of these thought experiments, Leibniz describes two universes which by Newton's own admission we could never tell apart – they are perfectly indiscernible. But by PII, this means that the two universes are actually one. So it follows that Newton's theory of absolute space is false. Another way to see the point is this. Newton's theory implies that there is a genuine difference between the universe being at one location in absolute space and it being shifted to a different location. But Leibniz points out that this difference would be totally undetectable, so long as every object shifts location by the same amount. But if no difference can be detected between two universes then they are indiscernible, and PII tells us that they are actually the same universe. So Newton's theory has a false consequence: it implies that there are two things when there is only one. The concept of absolute space thus conflicts with PII. The logic of Leibniz's second thought experiment is identical.

In effect, Leibniz is arguing that absolute space is an empty notion, because it makes no observational difference. If neither the location of objects in absolute space nor their velocity with respect to absolute space can ever be detected, why believe in absolute space at all? Leibniz is appealing to the quite reasonable principle that we should only postulate unobservable entities in science if their existence would make a difference that we can detect observationally.

But Newton thought he could show that absolute space *did* have observational effects. This is the point of his famous 'rotating bucket' argument. He asks us to imagine a bucket full of water, suspended by a rope through a hole attached to its base (Figure 12).

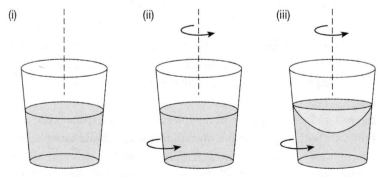

12. Newton's 'rotating bucket' experiment. In stage (i) bucket and water are at rest; in stage (ii) the bucket rotates relative to the water; in stage (iii) bucket and water rotate in tandem.

Initially the water is at rest relative to the bucket. Then the rope is twisted around a number of times and released. As it uncoils, the bucket starts rotating. At first the water in the bucket stays still, its surface flat; the bucket is then rotating relative to the water. But after a few moments the bucket imparts its motion to the water, and the water begins to rotate in tandem with the bucket; the bucket and the water are then at rest relative to each other again. Experience shows that the surface of the water then curves upwards at the sides, as the diagram indicates.

What is causing the surface of the water to rise?, Newton asks. Clearly it is something to do with the water's rotation. But rotation is a type of motion, and for Newton an object's motion is always relative to something else. So we must ask: relative to what is the water rotating? Not relative to the bucket, obviously, for the bucket and the water are rotating in tandem and are hence at relative rest. Newton argues that the water is rotating relative to absolute space, and that this is causing its surface to curve upwards. So absolute space does in fact have observational effects.

You may think there is an obvious gap in Newton's argument. Granted the water is not rotating relative to the bucket, but why conclude that it must be rotating relative to absolute space? The water is rotating relative to the person doing the experiment, and relative to the earth's surface, and relative to the fixed stars, so surely any of these might be causing its surface to rise? But Newton had a simple response to this move. Imagine a universe containing nothing except the rotating bucket. In such a universe, we cannot explain the water's curved surface by appealing to the water's rotation relative to other objects, for there are none, and as before the water is at rest relative to the bucket. Absolute space is the only thing left for the water to be rotating relative to. So we must believe in absolute space on pain of being unable to explain why the water's surface curves.

In effect, Newton is saying that although an object's position in absolute space and its velocity with respect to absolute space can never be detected, it *is* possible to tell when an object is *accelerating* with respect to absolute space. For when an object rotates then it is by definition accelerating, even if the rate of rotation is constant. This is because in physics, acceleration is defined as the rate of change of velocity, and velocity is speed *in a fixed direction*. Since rotating objects are constantly changing their direction of motion, it follows that their velocity is not constant, hence they are accelerating. The water's curved surface is just one example of what are called 'inertial effects' – effects produced by accelerated motion. Another example is the feeling of being pushed to the back of your seat that you get when an aeroplane takes off. The only possible explanation of inertial effects, Newton believed, is the acceleration of the object experiencing those effects with respect to absolute space. For in a universe containing only the accelerating object, absolute space is the only thing that the acceleration could be relative to.

Newton's argument is powerful but not conclusive. For how does Newton know that the water's surface *would* curve upwards, if the rotating bucket experiment was done in a universe containing no other material objects? Newton simply assumes that the inertial effects we find in this world would remain the same in a world bereft of any other matter. This is obviously quite a substantial assumption, and many people have questioned Newton's entitlement to it. So Newton's argument does not prove the existence of absolute space. Rather, it lays down a challenge to the defender of Leibniz to provide an alternative explanation of inertial effects.

Leibniz also faces the challenge of explaining the difference between absolute and relative motion without invoking absolute space. On this problem, Leibniz wrote that a body is in true or absolute motion 'when the immediate cause of the change is in the body itself'. Recall the case of the hang-glider and the observer on earth, both of whom are in motion relative to the other. To determine which is 'really' moving, Leibniz would say that we need to decide whether the immediate cause of the change (i.e. of the relative motion) is in the hang-glider, the observer, or both. This suggestion for how to distinguish absolute from relative motion avoids all reference to absolute space, but it is not very clear. Leibniz never properly explains what it *means* for the 'immediate cause of the change' to be in an object. But it may be that he intended to reject Newton's assumption that an object's motion, whether relative or absolute, can only be a fact about the object's relations to something else.

One of the intriguing things about the absolute/relational controversy is that it refuses to go away. Newton's account of space was intimately bound up with his physics, and Leibniz's views were a direct reaction to Newton's. So one might think that the advances in physics since the 17th century would have resolved the issue by now. But this has not happened. Although it was once widely held that Einstein's theory of relativity had decided the issue in favour of Leibniz, this view has increasingly come under attack in recent years. More than 300 years after the original Newton/Leibniz debate, the controversy rages on.

The problem of biological classification

Classifying, or sorting the objects one is studying into general kinds, plays a role in every science. Geologists classify rocks as igneous, sedimentary, or metamorphic, depending on how they were formed. Economists classify taxation systems as proportional, progressive, or regressive, depending on how unfair they are. The main function of classification is to convey information. If a chemist tells you that something is a metal, that tells you a lot about its likely behaviour. Classification raises some interesting philosophical issues. Mostly, these stem from the fact that any given set of objects can in principle be classified in many different ways. Chemists classify substances by their atomic number, yielding the periodic table of the elements. But they could equally classify substances by their

colour, or their smell, or their density. So how should we choose
between these alternative ways of classifying? Is there a 'correct'
way to classify? Or are all classification schemes ultimately
arbitrary? These questions take on a particular urgency in the
context of biological classification, or taxonomy, which will be our
concern here.

Biologists traditionally classify plants and organisms using the
Linnean system, named after the 18th-century Swedish naturalist
Carl Linnaeus (1707–1778) (Figure 13). The basic elements of the
Linnean system are straightforward, and familiar to many people.
First of all, individual organisms are assigned to a *species*. Each
species is then assigned to a *genus*, each genus to a *family*, each
family to an *order*, each order to a *class*, each class to a *phylum*,
and each phylum to a *kingdom*. Various intermediate ranks, such
as *subspecies*, *subfamily*, and *superfamily* are also recognized.
The species is the base taxonomic unit; genuses, families, orders,
and so on are known as 'higher taxa'. The standard Latin name
for a species indicates the genus to which the species belongs, but
no more. For example, you and I belong to *Homo sapiens*, the
only surviving species in the Homo genus. Two of the other
species in that genus are *Homo erectus* and *Homo habilis*, both
now extinct. The Homo genus belongs to the Hominid family,
which belongs to the Hominoid superfamily, which belongs to the
Primate order, which belongs to the Mammalian class, which
belongs to the Chordate phylum, which belongs to the Animal
kingdom.

Notice that the Linnean way of classifying organisms is
hierarchical: a number of species are nested in a single genus, a
number of genuses in a single family, a number of families in a
single order, and so on. So as we move upwards, we find fewer taxa
at each level. At the bottom there are literally millions of species,
but at the top there are just five kingdoms: Animals, Plants, Fungi,
Bacteria, and Protoctists (algae, seaweed, etc.). Not every
classification system in science is hierarchical. The periodic table in

CAROLI LINNÆI
Naturæ Curioſorum *Dioſcoridis Secundi*

SYSTEMA
NATURÆ

IN QUO

NATURÆ REGNA TRIA,
SECUNDUM.

CLASSES, ORDINES, GENERA, SPECIES,

SYSTEMATICE PROPONUNTUR,

Editio Secunda, Auctior.

STOCKHOLMIÆ
Apud GOTTFR. KIESEWETTER.
1740.

13. Linneaus' most famous book *Systema Naturae*, in which he presented his classification of plants, animals, and minerals.

chemistry is an example of a non-hierarchical classification. The different chemical elements are not arranged into more and more inclusive groupings, the way species are in the Linnean system. One important question we must face is *why* biological classification should be hierarchical.

The Linnean system served naturalists well for hundreds of years, and continues to be used today. In some ways this is surprising, since biological theories have changed greatly in that period. The cornerstone of modern biology is Darwin's theory of evolution, which says that contemporary species have descended from ancestral species; this theory contrasts with the older, biblically inspired view that each species was created separately by God. Darwin's *Origin of Species* was published in 1859, but it was not until the middle of the 20th century that biologists began to ask whether the theory of evolution should have any impact on the way organisms are classified. By the 1970s two rival taxonomic schools had emerged, offering competing answers to this question. According to *cladists*, biological classifications should try to reflect the evolutionary relationships between species, so knowledge of evolutionary history is indispensable for doing good taxonomy. According to *pheneticists*, this is not so: classification can and should be totally independent of evolutionary considerations. A third group, known as the *evolutionary taxonomists*, try to combine elements of both views.

To understand the dispute between cladists and pheneticists, we must divide the problem of biological classification into two. Firstly, there is the problem of how to sort organisms into species, known as the 'species problem'. This problem has by no means been solved, but in practice biologists are often able to agree about how to delimit species, though there are difficult cases. Broadly speaking, biologists assign organisms to the same species if they can interbreed with each other and to different species otherwise. Secondly, there is the problem of how to arrange a group of species into higher taxa, which obviously presumes a solution to the first problem. As it happens, cladists and pheneticists do often disagree about the species problem, but their dispute primarily concerns higher taxa. So for the moment, we ignore the species problem – we assume that organisms have been allocated to species in a satisfactory way. The question is: where do we go from there? What principles do we use to classify these species into higher taxa?

To focus the issue, consider the following example. Humans, chimpanzees, gorillas, bonobos, orangutans, and gibbons are usually classed together as members of the Hominoid superfamily. But baboons are not counted as Hominoids. Why is this? What is the justification for placing humans, chimps, gorillas, etc. in a group that doesn't also contain baboons? According to pheneticists, the answer is that the former all have a number of features that baboons do not, for example the lack of a tail. On this view, taxonomic groupings should be based on *similarity* – they should bring together species that are similar to each other in important ways and leave out ones that are dissimilar. Intuitively, this is a reasonable view. For it fits neatly with the idea that the purpose of classification is to convey information. If taxonomic groups are based on similarity, then being told which group a particular organism belongs to will tell you a lot about its likely characteristics. If you are told that a given organism belongs to the Hominoid superfamily, you will know that it doesn't have a tail. Furthermore, many of the groups recognized by traditional taxonomy do seem to be similarity-based. To take an obvious example, plants all share a number of features that animals lack, so placing all the plants in one kingdom and all the animals in another makes good sense from the phenetic point of view.

However, cladists insist that similarity should count for nothing in classification. Rather what matters are the evolutionary relationships between species – known as their *phylogenetic* relations. Cladists agree that the baboons should be excluded from the group that contains humans, chimps, gorillas, etc. But the justification for this has got nothing to do with the similarities and dissimilarities between the species. The point is rather that the Hominoid species are more closely related to each other than are any of them to the baboons. What exactly does this mean? It means that all of the Hominoid species share a common ancestor that is not an ancestor of the baboons. Notice that this does *not* mean that the Hominoid species and the baboons have no common ancestor at all. On the contrary, any two species have a common ancestor if you go back far enough in evolutionary time – for all life on earth is presumed to have a single origin. The point is rather that the common ancestor of the Hominoid species and the baboons is also an ancestor of many other species, for example the various macaque species. So cladists argue that any taxonomic group that contains the Hominoid species and the baboons must also contain these other species. No taxonomic group can contain *just* the Hominoid species and the baboons.

The key cladistic idea is that all taxonomic groups, be they genuses, families, superfamilies, or whatever, must be *monophyletic*. A monophyletic group is one that contains an ancestral species and all of its descendants, but no-one else. Monophyletic groups come in various sizes. At one extreme, all species that have ever existed form a monophyletic group, presuming life on earth only originated once. At the other extreme, there can be monophyletic groups of just two species – if they are the only descendants of a common ancestor. The group that contains just the Hominoid species and the baboons is not monophyletic, for as we saw, the common ancestor of the Hominoid species and the baboons is also ancestral to the macaques. So it is not a genuine taxonomic group, according to cladists. Groups that are not monophyletic are not permitted in cladistic taxonomy, irrespective of how similar their members may be. For cladists regard such groupings as wholly artificial, by contrast with 'natural' monophyletic groups.

The concept of monophyly is easily understood graphically. Consider the diagram below – known as a *cladogram* – which shows the phylogenetic relationships between six contemporary species,

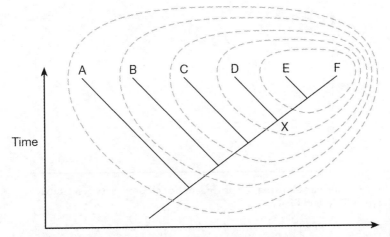

14. Cladogram showing the phylogenetic relations between six contemporary species.

A–F (Figure 14). All six species have a common ancestor if we go back far enough in time, but some are more closely related than others. Species E and F have a very recent common ancestor – for their branches intersect in the quite recent past. By contrast, species A split off from the rest of the lineage a long time ago. Now consider the group {D, E, F}. This is a monophyletic group, since it contains all and only the descendants of an ancestral species (not named), which split into two at the node marked 'x'. The group {C, D, E, F} is likewise monophyletic, as is the group {B, C, D, E, F}. But the group {B, C, D, F} is not monophyletic. This is because the common ancestor of these four species is also an ancestor of species E. All the monophyletic groups in the diagram have been ringed; any other group of species is not monophyletic.

The dispute between cladists and pheneticists is by no means purely academic – there are many real cases where they disagree. One well-known example concerns the class Reptilia, or the reptiles. Traditional Linnean taxonomy counts lizards and crocodiles as members of Reptilia, but excludes birds, which are placed in a separate class called Aves. Pheneticists agree with this traditional

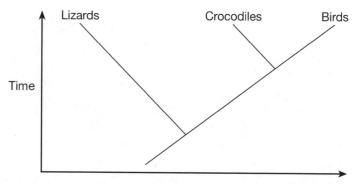

15. Cladogram showing the phylogenetic relations between lizards, crocodiles, and birds.

classification, for birds have their own unique anatomy and physiology, which is quite different from that of lizards, crocodiles, and other reptiles. But cladists maintain that Reptilia is not a genuine taxonomic group at all, for it is not monophyletic. As the cladogram above shows, the common ancestor of the lizards and the crocodiles is also an ancestor of the birds; so placing lizards and crocodiles together in a group that excludes birds violates the requirement of monophyly (Figure 15). Cladists therefore recommend that traditional taxonomic practice be abandoned: biologists should not talk about Reptilia at all, for it is an artificial not a natural group. This is quite a radical recommendation; even biologists sympathetic to the spirit of cladism are often reluctant to abandon the traditional taxonomic categories that have served naturalists well for centuries.

Cladists argue that their way of classifying is 'objective' while that of the pheneticists is not. There is certainly some truth in this charge. For pheneticists base their classifications on the similarities between species, and judgements of similarity are invariably partly subjective. Any two species are going to be similar to each other in some respects, but not in others. For example, two species of insect might be anatomically quite similar, but very diverse in their feeding habits. So which 'respects' do we single out, in order to make judgements of similarity? Pheneticists hoped to avoid this problem by defining a measure of 'overall similarity', which would take into account all of a species' characteristics, thus permitting fully objective classifications to be constructed. But though this idea sounds nice, it did not work, not least because there is no obvious way to count characteristics. Most people today believe that the very idea of 'overall similarity' is philosophically suspect. Phenetic classifications do exist, and are used in practice, but they are not fully objective. Different similarity judgements lead to different phenetic classifications, and there is no obvious way to choose between them.

Cladism faces its own set of problems. The most serious problem is that in order to construct a classification according to cladistic principles, we need to discover the phylogenetic relations between the species we are trying to classify, and this is very far from easy. These relations are obviously not discoverable just by looking at the species – they have to be inferred. A variety of techniques for inferring phylogenetic relations have been developed, but they are not fool-proof. Indeed, as more and more evidence from molecular genetics emerges, hypotheses about the phylogenetic relations between species get overturned rapidly. So actually putting cladistic ideas into practice is not easy. It is all very well to be told that only monophyletic groups of species are allowed in taxonomy, but this is of limited use unless one knows whether a given group *is* monophyletic or not. In effect, cladistic classifications constitute hypotheses about the phylogenetic relations between species, and are thus inherently conjectural. Pheneticists object that classification should not be theory-laden in this way. They maintain that taxonomy should be prior to, not dependent on, conjectures about evolutionary history.

Despite the difficulty of putting cladism into practice, and despite the fact the cladists often recommend quite radical revisions of traditional taxonomic categories, more and more biologists are coming round to the cladistic viewpoint. This is mainly because cladism is free of ambiguity in a way that phenetic and other approaches are not – its taxonomic principles are perfectly clear, even if they are hard to implement. And there is something quite intuitive about the idea that monophyletic groups of species are 'natural units', while other groups are not. Furthermore, cladism provides a genuine rationale for why biological classification should be hierarchical. As Figure 15 above indicates, monophyletic groups are always nested inside each other, so if the requirement of monophyly is rigidly followed the resulting classification will automatically be hierarchical. Classifying on the basis of similarity can also yield a hierarchical classification; but pheneticists have no comparable justification for *why* biological classification should be hierarchical. It is quite striking that naturalists have been classifying living organisms hierarchically for hundreds of years, but the true rationale for doing so has only recently become clear.

Is the mind modular?

One of the central jobs of psychology is to understand how human beings manage to perform the cognitive tasks they do. By 'cognitive tasks' we do not just mean things like solving crossword puzzles, but also more mundane tasks like crossing the road safely, understanding what other people say, recognizing other people's faces, checking one's change in a shop, and so on. There is no denying that humans are very good at many of these tasks – so good, indeed, that we often do them very fast, with little if any conscious thought. To appreciate just how remarkable this is, consider the fact that no robot has ever been designed that behaves even remotely like a human being in a real-life situation, despite considerable effort and expense. No robot can solve a crossword, or engage in a conversation, with anything like the facility the average human being can. Somehow or other, we humans are capable of performing complex cognitive tasks with minimal effort. Trying to understand how this could be is the central explanatory problem of the discipline known as cognitive psychology.

Our focus is an old but ongoing debate among cognitive psychologists concerning the architecture of the human mind. According to one view, the human mind is a 'general-purpose problem-solver'. This means that the mind contains a set of general problem-solving skills, or 'general intelligence', which it applies to an indefinitely large number of different tasks. So one and the same set of cognitive capacities is employed, whether the human is trying to count marbles, decide which restaurant to eat in, or learn a foreign language – these tasks represent different applications of the human's general intelligence. According to a rival view, the human mind contains a number of specialized subsystems or modules, each of which is designed for performing a very limited range of tasks and cannot do anything else (Figure 16). This is known as the *modularity of mind* hypothesis. So, for example, it is widely believed that there is a special module for language acquisition, a view deriving from the work of the linguist Noam Chomsky. Chomsky insisted that a child does not learn to speak by overhearing adult conversation and then using his 'general intelligence' to figure out the rules of the language being spoken; rather, there is a distinct 'language acquisition device' in every human child which operates automatically, and whose sole function is to enable him or her to learn a language, given appropriate prompting. Chomsky provided an array of impressive evidence for this claim – including, for example, the fact that even those with very low 'general intelligence' can often learn to speak perfectly well.

Some of the most compelling evidence for the modularity hypothesis comes from studies of patients with brain damage, known as 'deficit studies'. If the human mind is a general-purpose problem-solver, we would expect damage to the brain to affect all cognitive capacities more or less equally. But this is not what we find. On the contrary, brain damage often impairs some cognitive capacities but leaves others untouched. For example, damage to a part of the brain known as Wernicke's area leaves patients unable to understand speech, though they are still able to produce fluent,

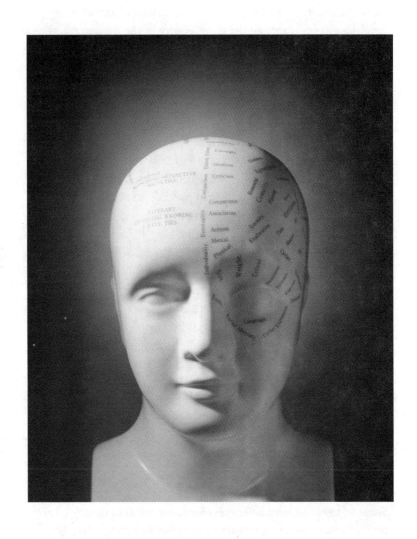

16. A hypothetical representation of a modular mind.

grammatical sentences. This strongly suggests that there are separate modules for sentence production and comprehension – for that would explain why loss of the latter capacity does not entail loss of the former. Other brain-damaged patients lose their long-term memory (amnesia), but their short-term memory and their ability to speak and understand are entirely unimpaired. Again, this seems to speak in favour of modularity and against the view of the mind as a general-purpose problem-solver.

Though compelling, neuropsychological evidence of this sort does not settle the modularity issue once and for all. For one thing, the evidence is relatively sparse – we obviously cannot damage people's brains at will just to see how their cognitive capacities are affected. In addition, there are serious disagreements about how the data should be interpreted, as is usual in science. Some people argue that the observed pattern of cognitive impairment in brain-damaged patients does not imply that the mind is modular. Even if the mind *were* a general-purpose problem-solver, that is non-modular, it is still possible that distinct cognitive capacities might be differentially affected by brain damage, they argue. So we cannot simply 'read off' the architecture of the mind from deficit studies, they maintain; at best, the latter provide fallible evidence for the former.

Much of the recent interest in modularity is due to the work of Jerry Fodor, an influential American philosopher and psychologist. In 1983 Fodor published a book called *The Modularity of Mind* which contained both a very clear account of what exactly a module is, and some interesting hypotheses about which cognitive capacities are modular and which not. Fodor argued that mental modules have a number of distinguishing features, of which the following three are the most important: (i) they are *domain-specific*, (ii) their operation is *mandatory*, and (iii) they are *informationally encapsulated*. Non-modular cognitive systems possess none of these features. Fodor then argued that the human mind is partly, though not wholly, modular: we solve some cognitive tasks using specialized modules, others using our 'general intelligence'.

To say that a cognitive system is domain-specific is to say that it is specialized: it performs a limited, precisely circumscribed set of tasks. Chomsky's postulated 'language acquisition device' is a good example of a domain-specific system. The sole function of this device is to enable the child to learn language – it doesn't help the child learn to play chess, or to count, or to do anything else. So the device simply ignores non-linguistic inputs. To say that a cognitive system is mandatory is to say that we cannot choose whether or not to put the system into operation. The perception of language provides a good example. If you hear a sentence uttered in a language you know, you cannot help but hear it as the utterance of a sentence. If someone asked you to hear the sentence as 'pure noise' you could not obey them however hard you tried. Fodor points out that not all cognitive processes are mandatory in this way. *Thinking* clearly is not. If someone asked you to think of the scariest moment in your life, or to think of what you would most like to do if you won the lottery, you clearly could obey their instructions. So thinking and language perception are quite different in this regard.

What about information encapsulation, the third and most crucial feature of mental modules? This notion is best illustrated by an example. Look at the two lines in Figure 17.

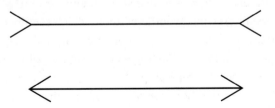

17. **The Müller-Lyer illusion. The horizontal lines are equal in length, but the top one looks longer.**

To most people, the top line looks slightly longer than the bottom one. But in fact this is an optical illusion, known as the Müller-Lyer illusion. The lines are actually equal in length. Various explanations have been suggested for why the top line looks longer, but they need not concern us here. The crucial point is this: the lines continue to look unequal in length, *even when you know it's an optical illusion*. According to Fodor, this simple fact has important implications for understanding the architecture of the mind. For it shows that the information that the two lines are equal in length is stored in a region of the cognitive mind to which our perceptual mechanisms do not have access. This means that our perceptual mechanisms are informationally encapsulated – they do not have access to all of the information we possess. If visual perception were not informationally encapsulated in this way, but could make use of all the information stored in the mind, then the illusion would disappear as soon as you were told that the lines were actually equal in length.

Another possible example of information encapsulation comes from the phenomenon of human phobias. Take, for example, odiophobia, or fear of snakes. This phobia is quite widespread in humans, and also in many other primate species. This is easily understood, for snakes are very dangerous to primates, so an instinctive fear of snakes could easily have evolved by natural selection. But whatever the explanation for why we are so scared of snakes, the crucial point is this. Even if you know that a particular snake isn't dangerous, for example because you've been told that its poison glands have been removed, you are still quite likely to be terrified of the snake and will not want to touch it. Of course, this sort of phobia can often be overcome by training, but that is a different matter. The relevant point is that the information that the snake isn't dangerous is inaccessible to the part of your mind that produces in you the reaction of fear when you see a snake. This suggests that there may be an inbuilt, informationally encapsulated 'fear of snakes' module in every human being.

You may wonder why the modularity of mind issue is at all philosophical. Surely it is just a question of empirical fact whether the mind is modular or not, albeit not an easy one to answer? In fact this suggestion is not quite right. One respect in which the modularity debate is philosophical concerns how we should count cognitive tasks and modules. Advocates of modularity hold that the mind contains specialized modules for performing different sorts of cognitive task; opponents of modularity deny this. But how do we decide whether two tasks are of the same sort, or of different sorts? Is facial recognition a single cognitive task or is it comprised of two distinct cognitive tasks: recognizing male faces and recognizing female faces? Are doing long division and doing multiplication different cognitive tasks, or are they both part of the more general task of doing arithmetic? Questions of this sort are conceptual or philosophical, rather than straightforwardly empirical, and they are potentially crucial to the modularity debate. For suppose an opponent of modularity produces some experimental evidence to show that we use one and the same set of cognitive capacities to perform many different types of cognitive task. Her opponent might accept the experimental data, but argue that the cognitive tasks in questions are all of the *same* type, and thus that the data are perfectly compatible with modularity. So first appearances to the contrary notwithstanding, the modularity of mind debate is up to its neck in philosophical issues.

The most enthusiastic advocates of modularity believe that the mind is entirely composed of modules, but this view is not widely accepted. Fodor himself argues that perception and language are probably modular, while thought and reasoning are almost certainly not. To see why not, suppose you are sitting on a jury and are trying to decide whether to return a verdict of guilty or not guilty. How will you go about your task? One important issue you will consider is whether the defendant's story is logically consistent or not – is it free from contradiction? And you will probably ask yourself whether the available evidence is merely compatible with the defendant's guilt or whether it strongly supports it. Clearly, the reasoning skills you apply here – testing for logical consistency and assessing evidence – are *general* skills; they are not specifically designed for use in jury service. You use the same skills in many domains. So the cognitive capacities you bring to bear in deliberating the defendant's guilt are not domain-specific. Nor is their operation mandatory – you have to consciously consider whether the defendant is guilty, and can stop doing so whenever you want to, e.g. during the lunch break. Most important of all, there is no information encapsulation either. Your task is to decide whether the defendant is guilty *all things considered,* so you may have to draw on any of the background information that you possess, if you consider it relevant. For example, if the defendant twitched nervously under cross-examination and you believe that nervous twitching is invariably a sign of guilt, you will probably draw on this belief in reaching your verdict. So there is no store of information which is inaccessible to the cognitive mechanisms you employ to reach your verdict (though the judge may tell you to ignore certain things). In short, there is no module for deciding whether a defendant is guilty. You tackle this cognitive problem using your 'general intelligence'.

Fodor's thesis that the mind is partly though not wholly modular thus looks quite plausible. But exactly how many modules there are, and what precisely they do, are questions that cannot be answered given the current state of research. Fodor himself is quite pessimistic about the possibility of cognitive psychology ever explaining the workings of the human mind. He believes that only modular systems can be studied scientifically – non-modular systems, because they are not informationally encapsulated, are much more difficult to model. So according to Fodor the best research strategy for cognitive psychologists is to focus on perception and language, ignoring thinking and reasoning. But this aspect of Fodor's thought is very controversial. Not all psychologists agree with him about which bits of the mind are modular and which are not, and not all agree that only modular systems can be studied scientifically.

Chapter 7
Science and its critics

Many people take it for granted that science is a good thing, for obvious reasons. After all, science has given us electricity, safe drinking water, penicillin, contraception, air travel, and much more – all of which have undoubtedly benefited humanity. But despite these impressive contributions to human welfare, science is not without its critics. Some argue that society spends too much money on science at the expense of the arts; others hold that science has given us technological capabilities we would be better off without, such as the capacity to produce weapons of mass destruction (Figure 18). Certain feminists argue that science is

18. Scientific capabilities we would be better off without: a toxic mushroom cloud produced by an atomic explosion.

objectionable because it is inherently male-biased; those of religious persuasion often feel that science threatens their faith; and anthropologists have accused Western science of arrogance, on the grounds that it blithely assumes its superiority to the knowledge and beliefs of indigenous cultures around the world. This by no means exhausts the list of criticisms to which science has been subject, but in this chapter we confine our attention to three that are of particular philosophical interest.

Scientism

The words 'science' and 'scientific' have acquired a peculiar cachet in modern times. If someone accuses you of behaving 'unscientifically', they are almost certainly criticizing you. Scientific conduct is sensible, rational, and praiseworthy; unscientific conduct is foolish, irrational, and worthy of contempt. It is difficult to know why the label 'scientific' should have acquired these connotations, but it is probably something to do with the high status in which science is held in modern society. Society treats scientists as experts, whose opinions are regularly sought on matters of importance and for the most part accepted without question. Of course, everybody recognizes that scientists sometimes get it wrong – for example, scientific advisers to the British government in the 1990s declared that 'mad cow disease' posed no threat to humans, only to be proved tragically mistaken. But occasional hiccups of this sort tend not to shake the faith that the public place in science, nor the esteem in which scientists are held. In the West at least, scientists are viewed much as religious leaders used to be: possessors of specialized knowledge that is inaccessible to the laity.

'Scientism' is a pejorative label used by some philosophers to describe what they see as science-worship – the over-reverential attitude towards science found in many intellectual circles. Opponents of scientism argue that science is not the only valid form of intellectual endeavour, and not the uniquely privileged route to knowledge. They often stress that they are not anti-science *per se*; what they are opposed to is the privileged status accorded to science, particularly natural science, in modern society, and the assumption that the methods of science are necessarily applicable to every subject matter. So their aim is not to attack science but to put it in place – to show that science is simply one among equals, and to free other disciplines from the tyranny that science supposedly exerts over them.

Scientism is obviously quite a vague doctrine, and since the term is in effect one of abuse, almost nobody would admit to believing it. Nonetheless, something quite like science-worship is a genuine feature of the intellectual landscape. This is not necessarily a bad thing – perhaps science deserves to be worshipped. But it is certainly a real phenomenon. One field that is often accused of science-worship is contemporary Anglo-American philosophy (of which philosophy of science is just one branch). Traditionally, philosophy is regarded as a humanities subject, despite its close historical links to mathematics and science, and with good reason. For the questions that philosophy addresses include the nature of knowledge, of morality, of rationality, of human well-being, and more, none of which appear soluble by scientific methods. No branch of science tells us how we should lead our lives, what knowledge is, or what human happiness involves; these are quintessentially philosophical questions.

Despite the apparent impossibility of answering philosophical questions through science, quite a few contemporary philosophers do believe that science is the only legitimate path to knowledge. Questions that cannot be resolved by scientific means are not genuine questions at all, they hold. This view is often associated with the late Willard van Orman Quine, arguably the most important American philosopher of the 20th century. The grounds for the view lie in a doctrine called 'naturalism', which stresses that we human beings are part and parcel of the natural world, not something apart from it, as was once believed. Since science studies the whole of the natural world, surely it should be capable of revealing the complete truth about the human condition, leaving nothing left for philosophy? Adherents of this view sometimes add that science undeniably makes progress, while philosophy seems to discuss the same questions for centuries on end. On this conception, there is no such thing as distinctively philosophical knowledge, for all knowledge is scientific knowledge. In so far as there is a role for philosophy at all, it consists in 'clarifying scientific concepts' – clearing the brush so that scientists can get on with their work.

Not surprisingly, many philosophers reject this subordination of their discipline to science; this is one of the main sources of opposition to scientism. They argue that philosophical enquiry reveals truths about a realm that science cannot touch. Philosophical questions are incapable of being resolved by scientific means, but are none the worse for that: science is not the only path to the truth. Proponents of this view can allow that philosophy should aim to be *consistent* with the sciences, in the sense of not advancing claims that conflict with what science teaches us. And they can allow that the sciences deserve to be treated with great respect. What they reject is scientific imperialism – the idea that science is capable of answering all the important questions about man and his place in nature. Advocates of this position usually think of themselves as naturalists too. They do not normally hold that we humans are somehow outside the natural order, and so exempt from the scope of science. They allow that we are just another biological species, and that our bodies are ultimately composed of physical particles, like everything else in the universe. But they deny that this implies that scientific methods are appropriate for addressing every question of interest.

A similar issue arises regarding the relation between the natural sciences and the social sciences. Just as philosophers sometimes complain of 'science worship' in their discipline, so social scientists sometimes complain of 'natural science worship' in theirs. There is no denying that the natural sciences – physics, chemistry, biology, etc. – are in a more advanced state than the social sciences – economics, sociology, anthropology, etc. A number of people have wondered why this is so. It can hardly be because natural scientists are smarter than social scientists. One possible answer is that the *methods* of the natural sciences are superior to those of the social sciences. If this is correct, then what the social sciences need to do to catch up is to ape the methods of the natural sciences. And to some extent, this has actually happened. The increasing use of mathematics in the social sciences may be partly a result of this attitude. Physics made a great leap forward when Galileo took the step of applying mathematical language to the description of motion; so it is tempting to think that a comparable leap forward might be achievable in the social sciences, if a comparable way of 'mathematicizing' their subject matter can be found.

However, some social scientists strongly resist the suggestion that they should look up to the natural sciences in this way, just as some philosophers strongly resist the idea that they should look up to science as a whole. They argue that the methods of natural science are not necessarily appropriate for studying social phenomena. Why should the very same techniques that are useful in astronomy, for example, be equally useful for studying societies? Those who hold this view deny that the more advanced state of the natural sciences is attributable to the distinctive methods of enquiry they employ, and thus see no reason to extend those methods to the social sciences. They often point out that the social sciences are younger than the natural sciences, and that the complex nature of social phenomena makes successful social science very hard to do.

Neither the scientism issue nor the parallel issue about natural and social science is easy to resolve. In part, this is because it is far from clear what exactly the 'methods of science', or the 'methods of natural science', actually comprise – a point that is often overlooked by both sides in the debate. If we want to know whether the methods of science are applicable to every subject matter, or whether they are capable of answering every important question, we obviously need to know what exactly those methods *are*. But as we have seen in previous chapters, this is much less straightforward a question than it seems. Certainly we know some of the main features of scientific enquiry: induction, experimental testing, observation, theory construction, inference to the best explanation, and so on. But this list does not provide a precise definition of 'the scientific method'. Nor is it obvious that such a definition *could* be provided. Science changes greatly over time, so the assumption that there is a fixed, unchanging 'scientific method', used by all scientific disciplines at all times, is far from inevitable. But this assumption is implicit both in the claim that science is the one true path to knowledge *and* in the counter-claim that some questions cannot be answered by scientific methods. This suggests that, to some extent at least, the debate about scientism may rest on a false presupposition.

Science and religion

The tension between science and religion is old and well documented. Perhaps the best-known example is Galileo's clash with the Catholic Church. In 1633 the Inquisition forced Galileo to publicly recant his Copernican views, and condemned him to spend the last years of his life under house arrest in Florence. The Church objected to the Copernican theory because it contravened the Holy Scriptures, of course. In recent times, the most prominent science/religion clash has been the bitter dispute between Darwinists and creationists in the United States, which will be our focus here.

Theological opposition to Darwin's theory of evolution is nothing new. When the *Origin of Species* was published in 1859, it immediately attracted criticism from churchmen in England. The reason is obvious: Darwin's theory maintains that all current species, including humans, have descended from common ancestors over a long period of time. This theory clearly contradicts the Book of Genesis, which says that God created all living creatures over a period of six days. So the choice looks stark: either you believe Darwin or you believe the Bible, but not both. Nonetheless, many committed Darwinians have found ways to reconcile their Christian faith with their belief in evolution – including a number of eminent biologists. One way is simply not to think about the clash too much. Another, more intellectually honest way is to argue that the Book of Genesis should not be interpreted literally – it should be regarded as allegorical, or symbolic. For after all, Darwin's theory is quite compatible with the existence of God, and with many other tenets of Christianity. It is only the literal truth of the biblical story of creation that Darwinism rules out. So a suitably attenuated version of Christianity can be rendered compatible with Darwinism.

However, in the United States, particularly in the Southern states, many evangelical Protestants have been unwilling to bend their religious beliefs to fit scientific findings. They insist that the biblical account of creation is literally true, and that Darwin's theory of evolution is therefore completely wrong. This opinion is known as 'creationism', and is accepted by some 40% of the adult population in the US, a far greater proportion than in Britain and Europe. Creationism is a powerful political force, and has had considerable influence on the teaching of biology in American schools, much to the dismay of scientists. In the famous 'monkey trial' of the 1920s, a Tennessee school teacher was convicted of teaching evolution to his pupils, in violation of state law. (The law was finally overturned by the Supreme Court in 1967.) In part because of the monkey trial, the subject of evolution was omitted altogether from the biology curriculum in US high schools for many decades. Generations of American adults grew up knowing nothing of Darwin.

This situation began to change in the 1960s, sparking a fresh round of battles between creationists and Darwinists, and giving rise to the movement called 'creation science'. Creationists want high-school students to learn the biblical story of creation, exactly as it appears in the Book of Genesis. But the American constitution prohibits the teaching of religion in public schools. The concept of creation science was designed to circumvent this. Its inventors argued that the biblical account of creation provides a better scientific explanation of life on earth than Darwin's theory of evolution. So teaching biblical creation does not violate the constitutional ban, for it counts as science, not religion! Across the Deep South, demands were made for creation science to be taught in biology classes, and they were very often heeded. In 1981 the state of Arkansas passed a law calling for biology teachers to give 'equal time' to evolution and to creation science, and other states followed suit. Though the Arkansas law was ruled unconstitutional by a federal judge in 1982, the call for 'equal time' continues to be heard today. It is often presented as a fair compromise – faced with two conflicting sets of beliefs, what could be fairer than giving equal time to each? Opinion polls show that an overwhelming majority of American adults agree: they want creation science to be taught alongside evolution in the public schools.

However, virtually all professional biologists regard creation science as a sham – a dishonest and misguided attempt to promote religious beliefs under the guise of science, with extremely harmful educational consequences. To counter this opposition, creation scientists have put great effort into trying to undermine Darwinism. They argue that the evidence for Darwinism is very inconclusive, so Darwinism is not established fact but rather just a theory. In addition, they have focused on various internal disputes among Darwinians, and picked on a few incautious remarks by individual biologists, in an attempt to show that disagreeing with the theory of evolution is scientifically respectable. They conclude that since Darwinism is 'just a theory', students should be exposed to alternative theories too – such as the creationist one that God made the world in six days.

In a way, the creationists are perfectly correct that Darwinism is 'just a theory' and not proven fact. As we saw in Chapter 2, it is never possible to *prove* that a scientific theory is true, in the strict sense of proof, for the inference from data to theory is invariably non-deductive. But this is a general point – it has nothing to do with the theory of evolution *per se*. By the same token, we could argue that it is 'just a theory' that the earth goes round the sun, or that water is made of H_2O, or that unsupported objects tend to fall, so students should be presented with alternatives to each of these. But creation scientists do not argue this. They are not sceptical about science as a whole, but about the theory of evolution in particular. So if their position is to be defensible, it cannot simply turn on the point that our data doesn't guarantee the truth of Darwin's theory. For the same is true of every scientific theory, and indeed of most common-sense beliefs too.

To be fair to the creation scientists, they do offer arguments that are specific to the theory of evolution. One of their favourite arguments is that the fossil record is extremely patchy, particularly when it comes to the supposed ancestors of *Homo sapiens*. There is some truth in this charge. Evolutionists have long puzzled over the gaps in the fossil record. One persistent puzzle is why there are so few 'transition fossils' – fossils of creatures intermediate between two species. If later species evolved from earlier ones as Darwin's theory asserts, surely we would expect transition fossils to be very common? Creationists take puzzles of this sort to show that Darwin's theory is just wrong. But the creationist arguments are uncompelling, notwithstanding the real difficulties in understanding the fossil record. For fossils are not the only or even the main source of evidence for the theory of evolution, as creationists would know if they had read *The Origin of Species*. Comparative anatomy is another important source of evidence, as are embryology, biogeography, and genetics. Consider, for example, the fact that humans and chimpanzees share 98% of their DNA. This and thousands of similar facts make perfect sense if the theory of evolution is true, and thus constitute excellent evidence for the theory. Of course, creation scientists can explain such facts too. They can claim that God decided to make humans and chimpanzees genetically similar, for reasons of His own. But the possibility of giving 'explanations' of this sort really just points to the fact that Darwin's theory is not logically entailed by the data. As we have seen, the same is true of every scientific theory. The creationists have merely highlighted the general methodological point that data can always be explained in a multitude of ways. This point is true, but shows nothing special about Darwinism.

Though the arguments of the creation scientists are uniformly unsound, the creationist/Darwinist controversy does raise important questions concerning science education. How should the clash between science and faith be dealt with in a secular education system? Who should determine the content of high-school science classes? Should tax payers have a say in what gets taught in the schools they pay for? Should parents who don't want their children to be taught about evolution, or some other scientific matter, be overruled by the state? Public policy matters such as these normally receive little discussion, but the clash between Darwinists and creationists has brought them to prominence.

Is science value free?

Almost everybody would agree that scientific knowledge has sometimes been used for unethical ends – in the manufacture of nuclear, biological, and chemical weapons, for example. But cases such as these do not show that there is something ethically objectionable about scientific knowledge itself. It is the *use* to which that knowledge is put that is unethical. Indeed, many philosophers would say that it makes no sense to talk about science or scientific knowledge being ethical or unethical *per se*. For science is concerned with facts, and facts in themselves have no ethical significance. It is what we do with those facts that is right or wrong, moral or immoral. According to this view, science is essentially a *value-free* activity – its job is just to provide information about the world. What society chooses to do with that information is another matter.

Not all philosophers accept this picture of science as neutral with respect to matters of value, nor the underlying fact/value dichotomy on which it rests. Some argue that the ideal of value-neutrality is unattainable – scientific enquiry is invariably laden with value judgements. (This is analogous to the claim that all observation is theory-laden, discussed in Chapter 4. Indeed, the two claims are often found hand-in-hand.) One argument against the possibility of value-free science stems from the obvious fact that scientists have to choose what to study – not everything can be examined at once. So judgements about the relative importance of different possible objects of study will have to be made, and these are value judgements, in a weak sense. Another argument stems from the fact, with which you should now be familiar, that any set of data can in principle be explained in more than one way. A scientist's choice of theory will thus never be uniquely determined by his data. Some philosophers take this to show that values are inevitably involved in theory choice, and thus that science cannot possibly be value-free. A third argument is that scientific knowledge cannot be divorced from its intended applications in the way that value-neutrality would require. On this view, it is naïve to picture scientists as disinterestedly doing research for its own sake, without a thought for its practical applications. The fact that much scientific research today is funded by private enterprises, who obviously have vested commercial interests, lends some credence to this view.

Though interesting, these arguments are all somewhat abstract – they seek to show that science could not be value free as a matter of principle, rather than identifying actual cases of values intruding in science. But specific accusations of value-ladenness have also been made. One such case concerns the discipline called human sociobiology, which generated considerable controversy in the 1970s and 1980s. Human sociobiology is the attempt to apply principles of Darwinian theory to human behaviour. At first blush this project sounds perfectly reasonable. For humans are just another species of animal, and biologists agree that Darwinian theory can explain a lot of animal behaviour. For example, there is an obvious Darwinian explanation for why mice usually run away when they see cats. In the past, mice that did not behave this way tended to leave fewer offspring than ones that did, for they got eaten; assuming that the behaviour was genetically based, and thus transmitted from parents to offspring, over a number of generations it would have spread through the population. This explains why mice today run away from cats. Explanations of this sort are known as 'Darwinian' or 'adaptationist' explanations.

Human sociobiologists (henceforth simply 'sociobiologists') believe that many behavioural traits in humans can be given adaptationist explanations. One of their favourite examples is incest-avoidance. Incest – or sexual relations between members of the same family – is regarded as taboo in virtually every human society, and subject to legal and moral sanctions in most. This fact is quite striking, given that sexual mores are otherwise quite diverse across human societies. Why the prohibition on incest? Sociobiologists offer the following explanation. Children born of incestuous relationships often have serious genetic defects. So in the past, those who practised incest would have tended to leave fewer viable offspring than those who didn't. Assuming that the incest-avoiding behaviour was genetically based, and thus transmitted from parents to their offspring, over a number of generations it would have spread through the population. This explains why incest is so rarely found in human societies today.

Understandably enough, many people feel uneasy with this sort of explanation. For, in effect, sociobiologists are saying that we are genetically pre-programmed to avoid incest. This conflicts with the common-sense view that we avoid incest because we have been taught that it is wrong, i.e. that our behaviour has a cultural rather than a biological explanation. And incest-avoidance is actually one of the least controversial examples. Other behaviours for which sociobiologists offer adaptationist explanations include rape, aggression, xenophobia, and male promiscuity. In each case, their argument is the same: individuals who engaged in the behaviour

131

out-reproduced individuals who didn't, and the behaviour was genetically based, hence transmitted from parents to their offspring. Of course, not all humans are aggressive, xenophobic, or engage in rape. But this does not show that the sociobiologists are wrong. For their argument only requires that these behaviours have a genetic component, i.e. that there is some gene or genes which increases the probability that its carriers will engage in the behaviours. This is much weaker than saying that the behaviours are totally genetically determined, which is almost certainly false. In other words, the sociobiological story is meant to explain why there is a *disposition* among humans to be aggressive, xenophobic, and to rape – even if such dispositions are infrequently manifested. So the fact that aggression, xenophobia, and rape are (thankfully) quite rare does not in itself prove the sociobiologists wrong.

Sociobiology attracted strong criticism from a wide range of scholars. Some of this was strictly scientific. Critics pointed out that sociobiological hypotheses were extremely hard to test, and should thus be viewed as interesting conjectures, not established truths. But others objected more fundamentally, claiming that the whole sociobiological research programme was ideologically suspect. They saw it as an attempt to justify or excuse anti-social behaviour, usually by men. By arguing that rape, for example, has a genetic component, sociobiologists were implying that it was 'natural' and thus that rapists were not really responsible for their actions – they were simply obeying their genetic impulses. 'How can we blame rapists, if their genes are responsible for their behaviour?', the sociobiologists seemed to be saying. Sociobiological explanations of xenophobia and male promiscuity were regarded as equally pernicious. They seemed to imply that phenomena such as racism and marital infidelity, which most people regard as undesirable, were natural and inevitable – the product of our genetic heritage. In short, critics charged that sociobiology was a value-laden science, and the values it was laden with were very dubious. Perhaps unsurprisingly, these critics included many feminists and social scientists.

One possible response to this charge is to insist on the distinction between facts and values. Take the case of rape. Presumably, either there is a gene which disposes men to rape and which spread by natural selection, or there is not. It is a question of pure scientific fact, though not an easy one to answer. But facts are one thing, values another. Even if there is such a gene, that does not make rape excusable or acceptable. Nor does it make rapists any the less responsible for their actions, for nobody thinks such a gene would literally *force* men to rape. At most, the gene might predispose men to rape, but innate predispositions can be overcome by cultural training, and everybody is taught that rape is wrong. The same applies to xenophobia, aggression, and promiscuity. Even if sociobiological explanations of these behaviours are correct, this has no implications for how we should run society, or for any other political or ethical matters. Ethics cannot be deduced from science. So there is nothing ideologically suspect about sociobiology. Like all sciences, it is simply trying to tell us the facts about the world. Sometimes the facts are disturbing, but we must learn to live with them.

If this response is correct, it means we should sharply distinguish the 'scientific' objections to sociobiology from the 'ideological' objections. Reasonable though this sounds, there is one point it doesn't address: advocates of sociobiology have tended to be politically right-wing, while its critics have tended to come from the political left. There are many exceptions to this generalization, especially to the first half of it, but few would deny the trend altogether. If sociobiology is simply an impartial enquiry into the facts, what explains the trend? Why should there be any correlation at all between political opinions and attitudes towards sociobiology? This is a tricky question to answer. For though some sociobiologists may have had hidden political agendas, and though some of sociobiology's critics have had opposing agendas of their own, the correlation extends even to those who debate the issue in apparently scientific terms. This suggests, though does not prove, that the 'ideological' and 'scientific' issues may not be quite so easy to separate after all. So the question of whether sociobiology is a value-free science is less easy to answer than might have been supposed.

To conclude, it is inevitable that an enterprise such as science, which occupies so pivotal a role in modern society and commands so much public money, should find itself subject to criticism from a variety of sources. It is also a good thing, for uncritical acceptance of everything that scientists say and do would be both unhealthy and dogmatic. It is safe to predict that science in the 21st century, through its technological applications, will impact on everyday life to an even greater extent than it has already. So the question 'is science a good thing?' will become yet more pressing. Philosophical reflection may not produce a final, unequivocal answer to this question, but it can help to isolate the key issues and encourage a rational, balanced discussion of them.

Further reading

Chapter 1

A. Rupert Hall, *The Revolution in Science 1500–1750* (Longman, 1983) contains a good account of the scientific revolution. Detailed treatment of particular topics in the history of science can be found in R. C. Olby, G. N. Cantor, J. R. R. Christie, and M. J. S. Hodge (eds.), *Companion to the History of Modern Science* (Routledge, 1990). There are many good introductions to philosophy of science. Two recent ones include Alexander Rosenberg, *The Philosophy of Science* (Routledge, 2000) and Barry Gower, *Scientific Method* (Routledge, 1997). Martin Curd and J. A. Cover (eds.), *Philosophy of Science: The Central Issues* (W.W. Norton, 1998) contains readings on all the main issues in philosophy of science, with extensive commentaries by the editors. Karl Popper's attempt to demarcate science from pseudo-science can be found in his *Conjectures and Refutations* (Routledge, 1963). A good discussion of Popper's demarcation criterion is Donald Gillies, *Philosophy of Science in the 20th Century* (Blackwell, Part IV, 1993). Anthony O'Hear, *Karl Popper* (Routledge, 1980) is a general introduction to Popper's philosophical views.

Chapter 2

Wesley Salmon, *The Foundations of Scientific Inference* (University of Pittsburgh Press, 1967) contains a very clear discussion of all the issues raised in this chapter. Hume's original argument can be found in Book IV, section 4 of his *Enquiry Concerning Human Understanding*, ed.

L. A. Selby-Bigge (Clarendon Press, 1966). Strawson's article is in Richard Swinburne (ed.), *The Justification of Induction* (Oxford University Press, 1974); the other papers in this volume are also of interest. Gilbert Harman's paper on IBE is 'The Inference to the Best Explanation', *Philosophical Review* 1965 (74), pp. 88–95. Peter Lipton, *Inference to the Best Explanation* (Routledge, 1991), is a book-length treatment of the topic. Popper's attempted solution of the problem of induction is in *The Logic of Scientific Discovery* (Basic Books, 1959); the relevant section is reprinted in M. Curd and J. Cover (eds.), *Philosophy of Science* (W.W. Norton, 1998), pp. 426–32. A good critique of Popper is Wesley Salmon's 'Rational Prediction', also reprinted in Curd and Cover (eds.), pp. 433–44. The various interpretations of probability are discussed in Donald Gillies, *Philosophical Theories of Probability* (Routledge, 2000) and in Brian Skryms, *Choice and Chance* (Wadsworth, 1986).

Chapter 3

Hempel's original presentation of the covering law model can be found in his *Aspects of Scientific Explanation* (Free Press, 1965, essay 12). Wesley Salmon, *Four Decades of Scientific Explanation* (University of Minnesota Press, 1989) is a very useful account of the debate instigated by Hempel's work. Two collections of papers on scientific explanation are Joseph Pitt (ed.), *Theories of Explanation* (Oxford University Press, 1988) and David-Hillel Ruben (ed.), *Explanation* (Oxford University Press, 1993). The suggestion that consciousness can never be explained scientifically is defended by Colin McGinn, *Problems of Consciousness* (Blackwell, 1991); for discussion, see Martin Davies 'The Philosophy of Mind' in A. C. Grayling (ed.), *Philosophy: A Guide Through the Subject* (Oxford University Press, 1995) and Jaegwon Kim, *Philosophy of Mind* (Westview Press, 1993, chapter 7). The idea that multiple realization accounts for the autonomy of the higher-level sciences is developed in a difficult paper by Jerry Fodor, 'Special Sciences', *Synthese* 28, pp. 77–115. For more on the important topic of reductionism, see the papers in section 8 of M. Curd and J. Cover (eds.), *Philosophy of Science* (W.W. Norton, 1998) and the editors' commentary.

Chapter 4

Jarrett Leplin (ed.), *Scientific Realism* (University of California Press, 1984) is an important collection of papers on the realism/anti-realism debate. A recent book-length defence of realism is Stathis Psillos, *Scientific Realism: How Science Tracks Truth* (Routledge, 1999). Grover Maxwell's paper 'The Ontological Status of Theoretical Entities' is reprinted in M. Curd and J. Cover (eds.), *Philosophy of Science* (W.W. Norton, 1998), pp. 1052–63. Bas van Fraassen's very influential defence of anti-realism is in *The Scientific Image* (Oxford University Press, 1980). Critical discussions of van Fraassen's work, with replies by van Fraassen, can be found in C. Hooker and P. Churchland (eds.), *Images of Science* (University of Chicago Press, 1985). The argument that scientific realism conflicts with the historical record is developed by Larry Laudan in 'A Confutation of Convergent Realism', *Philosophy of Science* 1981 (48), pp. 19–48, reprinted in Leplin (ed.), *Scientific Realism*. The 'no miracles' argument was originally developed by Hilary Putnam; see his *Mathematics, Matter and Method* (Cambridge University Press, 1975), pp. 69ff. Larry Laudan's 'Demystifying Underdetermination' in M. Curd and J. Cover (eds.), *Philosophy of Science* (W.W. Norton, 1998), pp. 320–53, is a good discussion of the concept of underdetermination.

Chapter 5

Important papers by the original logical positivists can be found in H. Feigl and M. Brodbeck (eds.), *Readings in the Philosophy of Science* (Appleton-Century-Croft, 1953). Thomas Kuhn, *The Structure of Scientific Revolutions* (University of Chicago Press, 1963) is for the most part very readable; all post-1970 editions contain Kuhn's Postscript. Kuhn's later thoughts, and his reflections on the debate sparked by his book, can be found in 'Objectivity, Value Judgment and Theory Choice' in his *The Essential Tension* (University of Chicago Press, 1977), and *The Road Since Structure* (University of Chicago Press, 2000). Two recent book-length discussions of Kuhn's work are Paul Hoyningen-Heune, *Reconstructing Scientific Revolutions: Thomas Kuhn's Philosophy of Science* (University of Chicago Press, 1993) and Alexander Bird, *Thomas Kuhn* (Princeton University Press, 2001). Paul Horwich (ed.), *World Changes* (MIT Press, 1993) contains discussions of Kuhn's work by well-known historians and philosophers of science, with comments by Kuhn himself.

Chapter 6

The original debate between Leibniz and Newton consists of five papers by Leibniz and five replies by Samuel Clarke, Newton's spokesman. These are reprinted in H. Alexander (ed.), *The Leibniz-Clarke Correspondence* (Manchester University Press, 1956). Good discussions can be found in Nick Huggett (ed.), *Space from Zeno to Einstein* (MIT Press, 1999) and Christopher Ray, *Time, Space and Philosophy* (Routledge, 1991). Biological classification is discussed from a philosophical viewpoint by Elliott Sober, *Philosophy of Biology* (Westview Press, 1993, chapter 7). A very detailed account of the clash between pheneticists and cladists is given by David Hull, *Science as a Process* (University of Chicago Press, 1988). Also useful is Ernst Mayr, 'Biological Classification: Towards a Synthesis of Opposing Methodologies' in E. Sober (ed.), *Conceptual Issues in Evolutionary Biology*, 2nd edn. (MIT Press, 1994). Jerry Fodor, *The Modularity of Mind* (MIT Press, 1983) is quite difficult but well worth the effort. Good discussions of the modularity issue can be found in Kim Sterelny, *The Representational Theory of Mind* (Blackwell, 1990) and J. L. Garfield, 'Modularity', in S. Guttenplan (ed.), *A Companion to the Philosophy of Mind* (Blackwell, 1994).

Chapter 7

Tom Sorell, *Scientism* (Routledge, 1991) contains a detailed discussion of the concept of scientism. The issue of whether the methods of natural science are applicable to social science is discussed by Alexander Rosenberg, *Philosophy of Social Science* (Clarendon Press, 1988) and David Papineau, *For Science in the Social Sciences* (Macmillan, 1978). The creationist/Darwinist controversy is examined in detail by Philip Kitcher, *Abusing Science: The Case Against Creation* (MIT Press, 1982). A typical piece of creationist writing is Duane Gish, *Evolution? The Fossils Say No!* (Creation Life Publishers, 1979). Good general discussions of the issue of value-ladenness include Larry Laudan, *Science and Values* (University of California Press, 1984) and Helen

Longino, *Science as Social Knowledge: Values and Objectivity in Scientific Inquiry* (Princeton University Press, 1990). The controversy over sociobiology was instigated by Edward O. Wilson, *Sociobiology* (Harvard University Press, 1975); also relevant is his *On Human Nature* (Bantam Books, 1978). A detailed and fair examination of the controversy is given by Philip Kitcher, *Vaulting Ambition: Sociobiology and the Quest for Human Nature* (MIT Press, 1985).

"牛津通识读本"已出书目

古典哲学的趣味
人生的意义
文学理论入门、
大众经济学
历史之源
设计,无处不在
生活中的心理学
政治的历史与边界
哲学的思与惑
资本主义
美国总统制
海德格尔
我们时代的伦理学
卡夫卡是谁
考古学的过去与未来
天文学简史
社会学的意识
康德
尼采
亚里士多德的世界
西方艺术新论
全球化面面观
简明逻辑学
法哲学:价值与事实
政治哲学与幸福根基
选择理论
后殖民主义与世界格局

福柯
缤纷的语言学
达达和超现实主义
佛学概论
维特根斯坦与哲学
科学哲学
印度哲学祛魅
克尔凯郭尔
科学革命
广告
数学
叔本华
笛卡尔
基督教神学
犹太人与犹太教
现代日本
罗兰·巴特
马基雅维里
全球经济史
进化
性存在
量子理论
牛顿新传
国际移民
哈贝马斯
医学伦理
黑格尔

地球
记忆
法律
中国文学
托克维尔
休谟
分子
法国大革命
民族主义
科幻作品
罗素
美国政党与选举
美国最高法院
纪录片
大萧条与罗斯福新政
领导力
无神论
罗马共和国
美国国会
民主
英格兰文学
现代主义
网络
自闭症
德里达
浪漫主义
批判理论

德国文学	儿童心理学	电影
戏剧	时装	俄罗斯文学
腐败	现代拉丁美洲文学	古典文学
医事法	卢梭	大数据
癌症	隐私	洛克
植物	电影音乐	幸福
法语文学	抑郁症	免疫系统
微观经济学	传染病	银行学
湖泊	希腊化时代	景观设计学
拜占庭	知识	神圣罗马帝国
司法心理学	环境伦理学	大流行病
发展	美国革命	亚历山大大帝
农业	元素周期表	气候
特洛伊战争	人口学	第二次世界大战
巴比伦尼亚	社会心理学	中世纪
河流	动物	工业革命
战争与技术	项目管理	传记
品牌学	美学	公共管理
数学简史	管理学	社会语言学
物理学	卫星	物质
行为经济学	国际法	学习
计算机科学	计算机	化学